Atlas of Cilia Bioengineering and Biocomputing

RIVER PUBLISHERS SERIES IN BIOMEDICAL ENGINEERING

Series Editor:

DINESH KANT KUMAR
RMIT University
Australia

Indexing: All books published in this series are submitted to the Web of Science Book Citation Index (BkCI), to CrossRef and to Google Scholar.

The "River Publishers Series in Biomedical Engineering" is a series of comprehensive academic and professional books which focus on the engineering and mathematics in medicine and biology. The series presents innovative experimental science and technological development in the biomedical field as well as clinical application of new developments.

Books published in the series include research monographs, edited volumes, handbooks and textbooks. The books provide professionals, researchers, educators, and advanced students in the field with an invaluable insight into the latest research and developments.

Topics covered in the series include, but are by no means restricted to the following:

- Biomedical engineering
- Biomedical physics and applied biophysics
- Bio-informatics
- Bio-metrics
- Bio-signals
- Medical Imaging

For a list of other books in this series, visit www.riverpublishers.com

Atlas of Cilia Bioengineering and Biocomputing

Editors

Richard Mayne

University of the West of England, UK

Jaap den Toonder

Technische Universiteit Eindhoven
The Netherlands

Published 2018 by River Publishers
River Publishers
Alsbjergvej 10, 9260 Gistrup, Denmark
www.riverpublishers.com

Distributed exclusively by Routledge
4 Park Square, Milton Park, Abingdon, Oxon OX14 4RN
605 Third Avenue, New York, NY 10158

First published in paperback 2024

Atlas of Cilia Bioengineering and Biocomputing / by Richard Mayne, Jaap den Toonder.

Routledge is an imprint of the Taylor & Francis Group, an informa business

ISBN: 978-87-7022-002-6 (hbk)
ISBN: 978-87-7004-383-0 (pbk)
ISBN: 978-1-003-33728-7 (ebk)

DOI: 10.1201/9781003337287

Contents

Editor Foreword

It is my opinion that science is no longer a set of disparate professions: possibly it never was. In order to understand the complexities of nature and attempt to exert control over them, it is not simply enough to be a 'biologist', 'engineer' or 'mathematician'.

This concept is amply exemplified in the study of cilia which are, arguably, one of the most intricate and specialised cellular organelles. Several hundred years after cilia were first described, we have still not fully characterised this incredible biological structure. By extension, we find several aspects of cilia difficult to model, experimentally manipulate and artificially reproduce. A clear trend is visible in contemporary research which demonstrates that those who address tenacious problems such as these are required to adapt to being expert in a great many fields: the biomedical scientist who addresses ciliary abnormalities in cystic fibrosis patients must be conversant in fluid mechanics in order to design treatments; the engineer who wants to design a cilia-inspired micromixing chip must aquanint themself with the complexities of cellular ultrastructure. This is a long-winded description of the concept of 'multidisciplinarity', which I have chosen to draw out here because the term is so exorbitantly overused but often poorly appreciated.

For every researcher who campaigns for multidisciplinary education and research, a multitude of university executives, government officials and research council policymakers will proudly repeat this call. Regardless, there is at the time of writing a distinct lack of investment in groundbreaking multidisciplinary education and research, principally because such projects are seen as high-risk and/or long-term investments.

Nevertheless, advances are being made where interested parties to educate, engage and collaborate with multidisciplinary fields. This is true of every demographic from layperson to student to Professor. The aim of this media atlas is to demonstrate the efforts, energy and talents of researchers from an extremely wide range of scientific and technical disciplines into a single multidisciplinary volume representing the 'state-of-the-art' in the specific field of cilia-based research. Noting my hesitance to sub-divide the sciences, the chapters presented here are arranged in three loose categories: those drawing on life sciences applications, those focussing on producing physical synthetic devices through engineering and those which focus on numerical applications through modelling/simulation.

All of the authors were given an extremely loose remit and were asked to concisely present developments in their research and/or a summary of their own particular field of expertise in as many words and in whatever style as they saw fit, providing that their contribution was lavishly illustrated. Furthermore, it was emphasised that all contributions should be as informative as possible, whilst still being comprehensible to non-experts. The rationale behind these criteria was twofold: firstly, in order to allow for the authors to express themselves in a creative, uninhibited fashion and secondly, to make the work presented appeal to as wide a readership as possible.

I would like to conclude this foreword by extending my thanks to all of the authors who contributed towards this volume: I believe that, thanks to their sterling efforts, the goal to make an accessible multidisciplinary text has been achieved. I must also extend my thanks to my co-editor Jaap and my PI Andy Adamatzky, who effectively acted as patron for this project.

Richard Mayne
April 2018

List of Contributors

Alexander Alexeev, *George W. Woodruff School of Mechanical Engineering, Georgia Institute of Technology, Atlanta, GA, USA, 30332*
E-mail: alexander.alexeev@me.gatech.edu

Anna C. Balazs, *Chemical Engineering Department, University of Pittsburgh, Pittsburgh, PA 15261, USA*
E-mail: balazs@pitt.edu

Benjamin Gorissen, *Department of Mechanical Engineering, KULeuven, Celestijnenlaan 300, 3001 Heverlee, Belgium*
E-mail: benjamin.gorissen@kuleuven.be

Dominiek Reynaerts, *Department of Mechanical Engineering, KULeuven, Celestijnenlaan 300, 3001 Heverlee, Belgium*
E-mail: dominiek.reynaerts@kuleuven.be

Edoardo Milana, *Department of Mechanical Engineering, KULeuven, Celestijnenlaan 300, 3001 Heverlee, Belgium*
E-mail: edoardo.milana@kuleuven.be

Gabrielle Wheway, *Centre for Research in Biosciences, Faculty of Health and Applied Sciences, University of the West of England, Bristol BS16 1QY, United Kingdom*
E-mail: Gabrielle.Wheway@uwe.ac.uk

James G. H. Whiting, *Unconventional computing Group, Frenchay Campus, University of the West of England, Bristol BS16 1QY, United Kingdom*
E-mail: James.Whiting@uwe.ac.uk

Jaap M. J. den Toonder, *Department of Mechanical Engineering and Institute of Complex Molecular Systems, Eindhoven University of Technology, Eindhoven, 5600 MB, The Netherlands*
E-mail: j.m.j.d.toonder@tue.nl

Michaël De Volder, *Department of Mechanical Engineering, KULeuven, Celestijnenlaan 300, 3001 Heverlee, Belgium*
Institute for Manufacturing, University of Cambridge, 17 Charles Babbage Road, Cambridge CB3 0FS, UK
E-mail: mfld2@cam.ac.uk

Nathan Banka, *Mechanical Engineering Department, University of Washington, Seattle, WA 98195-2600 USA*
E-mail: nbanka@uw.edu

Olga Kuksenok, *Materials Science and Engineering Department, Clemson University, Clemson, SC 29634, USA*
E-mail: okuksen@clemson.edu

Peter J. Hesketh, *George W. Woodruff School of Mechanical Engineering, Georgia Institute of Technology, Atlanta, GA, USA, 30332*
E-mail: peter.hesketh@me.gatech.edu

Patrick Onck, *Zernike Institute for Advanced Materials, University of Groningen, Groningen, 9747 AG, The Netherlands*
E-mail: p.r.onck@rug.nl

Richard Mayne, *Unconventional Computing Laboratory, University of the West of England, Bristol BS16 1QY, United Kingdom*
Email: Richard.Mayne@uwe.ac.uk

Santosh Devasia, *Mechanical Engineering Department, University of Washington, Seattle, WA 98195-2600 USA*
E-mail: devasia@uw.edu

Srinivas Hanasoge, *George W. Woodruff School of Mechanical Engineering, Georgia Institute of Technology, Atlanta, GA, USA, 30332*
E-mail: srinivasgh@gatech.edu

Shuaizhong Zhang, *Department of Mechanical Engineering and Institute of Complex Molecular Systems, Eindhoven University of Technology, Eindhoven, 5600 MB, The Netherlands*
E-mail: s.zhang1@tue.nl

Syed Khaderi, *Department of Mechanical and Aerospace Engineering, Indian Institute of Technology Hyderabad, Kandi, Hyderabad 502285, India*
E-mail: snk@iith.ac.in

Ya Liu, *Chemical Engineering Department, University of Pittsburgh, Pittsburgh, PA 15261, USA*
E-mail: yal65@pitt.edu

Ye Wang, *Department of Mechanical Engineering and Institute of Complex Molecular Systems, Eindhoven University of Technology, Eindhoven, 5600 MB, The Netherlands*
E-mail: Y.Wang2@tue.nl

List of Figures

List of Tables

List of Abbreviations

AAT	Acetylated alpha tubulin
AC	Alternating current
AM	Anitplectic Metachrony
ATCC	American type culture collection
ATP	Adenosine triphosphate
BZ	Belousov–Zhabotinsky
CA	Cellular automata
cAMP	Cyclic adenosine monophosphate
CCW	Counter-clockwise
CP	Concentrated particle (MAC)
CPU	Central processing unit
CRISPR	Clustered regularly interspersed short palindromic repeats
CW	Clockwise
DAPI	4',6-diamidino-2-phenylindole
DC	Direct current
DNA	Deoxyribonucleic acid
DPD	Dissipative Particle Dynamics
DSB	Double-strand breaks
ER	Endoplasmic reticulum
FGF	Fibroblast growth factor
GFP	Green fluorescent protein
gLSM	gel Lattice Spring Model
IFT	Intraflagellar transport
iPSCs	Induced pluripotent stem cells
LAP	Linearly aligned magnetic (MAC)
LBM	Lattice Boltzmann model
LCST	Lower critical solution temperature
LOC	Lab-on-a-chip
MAC	Magnetic artificial cilia
MB	Magnetic bead

McH	Protonated merocyanine form
MEMS	Micro-electrical mechanical systems
mRNA	Memory RNA
NoM	No Metachrony
OCT	Optical coherence tomography
PBA	Poly(butyl acrylate)
PCR	Polymerase chain reaction
PDMS	Polydimethylsiloxane
PIV	Particle Image Velocimetry
PNIPAAm	poly(N-isopropylacrylamide)
PTFE	Polytetrafluoroethylene
PTV	Particle tracking volcimetry
Re	Reynolds number
RNA	Ribonucleic acid
SDS	Sodium dodecyl sulphate
SEM	Scanning electron microscopy
shRNA	Short hairpin RNA
siRNA	Short interfering RNA
SM	Symplectic Metachrony
SP	Spiro form
TALENs	Transcription activator-like effector nucleases
TAP	Tandem affinity purification
TGF	Transforming growth factor
UV	Ultraviolet

PART I

Biology

1

Biological Preliminaries for Cilia Study

Richard Mayne[1] and Gabrielle Wheway[2]

[1]Unconventional Computing Laboratory, University of the West of England, Bristol BS16 1QY, United Kingdom
[2]Centre for Research in Biosciences, Faculty of Health and Applied Sciences, University of the West of England, Bristol BS16 1QY, United Kingdom
E-mail: Richard.Mayne@uwe.ac.uk; Gabrielle.Wheway@uwe.ac.uk

> "Those I saw, I could nowdiscern to be furnish't with very thin legs, which was very pleasant to behold."
>
> – Antoni van Leeuwenhoek, 1677

1.1 Introduction

Cilia (singular "cilium") were the first cellular organelle to be identified and were first described circa 1675 by the first microscopist, Antoni van Leeuwenhoek, who wrote of them as a component of an unknown flat "animalcule" (a pre-Linnean term for what we now call the protozoa); he isolated from rainwater what appeared as "incredibly thin feet, or little legs, which were moved very nimbly" [1]. Cilia were not named as such until 1786, however, by Otto Müller [2], although whether this was the first attribution of the term (which means "eyelash" in Latin) is debatable.

For centuries, the functions of cilia were attributed to motility, or more specifically as providing motive force to ciliated protozoa or otherwise creating fluid currents in adjacent media when a component of epithelia in multi-cellular organisms. In 1898, however, Zimmerman defined a second, "solitary" variety of cilium possessed by mammalian cells which he named the "central flagella" and described as "[a single] connecting thread continuing above the superficially-located body and extending freely into the lumen [of epithelial cells of the human seminal vesicle]" before proceeding to hypothesize that they had a sensory role [3]. This second variety of cilium was renamed in 1968 as the "primary cilium" and their role in sensing was confirmed [4], although their functions have only recently begun to be fully delineated.

This chapter aims to provide a brief introduction to the structure, functions, and key characteristics of both varieties of cilia, in order to bring readers of all disciplines and specialisms to a level of understanding sufficient to properly appreciate the research presented in consequent chapters.

1.2 Primary Cilia

Primary cilia are antenna-like organelles extending from the apical surface of eukaryotic cells. They have been described on a wide variety of cell types, and are found almost ubiquitously on epithelial cells. They were first described in 1898 by Zimmerman [3] who described them as primary because they appeared on the surface of the cells before the multiple, motile cilia developed. Unlike motile cilia, which are found on the surface of epithelial cells in great numbers, only one single primary cilium grows per cell.

Like motile cilia, primary cilia have a microtubule backbone, the axoneme, which is composed of a radial array of nine doublets of alpha and beta microtubules, enveloped by membrane continuous with the cell plasma membrane. Primary cilia lack the central pair of microtubules and the dynein arms ($9 + 0$ structure) possessed by motile cilia (see the following section), and thus have no motility. Their main function is in receiving and transducing signals from the extracellular environment, very much like an antenna [5].

These signaling functions are conferred by a diverse array of membrane-bound proteins in the cilium membrane, including receptors for signaling pathway ligands, flow-sensing stretch activated channels for detecting fluid flow, and photoreceptive pigment molecules for the detecting of light. No protein translation occurs within this cilium, and so all proteins must be trafficked into the cilium. Membrane proteins are added via vesicle fusion at the base of the cilium, soluble proteins are trafficked into and out of the cilium via a process called intraflagellar transport (IFT) [6]. Proteins are transported along the axoneme via kinesin and dynein motor proteins, in association with IFT proteins. Entry to and exit from the cilium are tightly regulated by the transition zone at the base of the cilium [7]. This area is so-called because it is the region where the gamma tubulin triplets of the basal body transition into the microtubule doublets of the ciliary axoneme. A large complex of proteins at the transition zone act as a ciliary gate, controlling the protein composition of the cilium.

The axoneme is anchored at its base by the basal body, a gamma tubulin structure derived from the mother centriole of the centrosome, the structure which nucleates the spindle during cell division [8]. Thus, formation of cilia is tightly linked to the cell cycle. In order for cells to divide, the cilium must be resorbed to release the centrioles for function in mitosis. Similarly, the cilium can only form when the cell is not undergoing mitosis. Primary cilia are characteristic of post-mitotic differentiated cells, but can also form in the S and G phases of the cell cycle [9].

Interest in primary cilia has exploded over the past few decades, since their discovery as organelles of central importance in Hedgehog signaling pathway transduction [10], a crucial pathway for cellular communication during development and disease. Since then, a role for primary cilia in virtually every cell signaling pathway has been discovered [11].

Gone are the days when the primary cilium was disregarded as a vestigial organelle with no motility and therefore complete redundancy. Primary cilia are now centers of much exciting research, of interest and importance to scientists interested in protein trafficking, microtubule dynamics, cell cycle, structural biology, development and cancer medicine, and systems biology. As a discrete organelle distinct from the rest of the cell, with its own transport system and proteome, the cilium represents a system of reduced complexity ideal for systems biology studies [12, 13].

In addition to the fascinating basic biology of the primary cilium, this organelle is a focus of medical research owing to its significance in human health and disease. Inherited genetic defects in proteins of the primary cilium are associated with a broad range of developmental conditions known as the ciliopathies [14]. An emerging area of study is the role of cilia in cancer development, as these organelles are tightly linked with control of the cell cycle and signaling pathways such as Hedgehog and Wnt, which are commonly dysregulated in cancers [15].

In order to understand the biology of primary cilia, and their contribution to human health and disease, we require robust systems for genetic manipulation of ciliated cells and organisms.

1.3 Motile Cilia

1.3.1 Structure

Due to the description of motile cilia predating that of primary cilia, they are much more widely known of and are usually referred to simply as "cilia", despite their being expressed by only a few metazoan cell types. As the name implies, motile cilia exist to generate motive force; in humans, motile cilia occur where epithelia are in contact with substances that need to be moved, such as in respiratory surfaces in the trachea and bronchioles (i.e., to move mucous) (Figure 1.1) and the linings of the fallopian tubes (for "sweeping" ova to the uterus). All such surfaces are polarized. In unicellular forms of life, an extremely diverse category of single-celled protozoa also possess cilia, known as the "ciliates" (historically called the "infusoria"). These organisms, of which there may be over 30,000 varieties, share markedly different morphologies and ciliary arrangements, but all share the same basic machinery (Figure 1.2) [16]. It is worth noting that cilia are almost identical in structure but not beating pattern to a homologous structure, the eukaryotic flagellum, although it is beyond the scope of this chapter to examine this further.

As was mentioned in the previous section, the ultrastructure[1] of the motile cilium diverges slightly from the primary cilium: in addition to the nine doublet array, they also possess a central pair of microtubules (9+2 structure) which are longer than the peripheral doublets (Figure 1.3). Furthermore, proteins called dyneins are also present in motile cilia which are their force-generating component. At rest, the cilium is said to exist in "rigor" state as dynenins bind the "A" component of a peripheral microtubule couplet perpendicularly to the "B" component of an adjacent couplet, thereby holding it rigid. In the presence of a molecule of ATP, the dynein arms de-couple from the adjacent B microtubule. The ATP is subsequently hydrolyzed which causes the dynein arm to bend and re-attach to the corresponding B couplet at an angle. The arm then re-orientates to its perpendicular state, causing a sliding motion in the corresponding doublet.

This motile activity is coupled to the central singlet microtubules by a central sheath, which is anchored to the peripheral microtubule doublets via proteins known as radial spokes (Figure 1.3). The radial spikes are thought to account for the rapid and smooth movement of the cilium as it extends through an entire beat [18].

[1]For an in-depth review of contemporary hypotheses on the exact molecular mechanisms of ciliary beating, we refer the reader to [17].

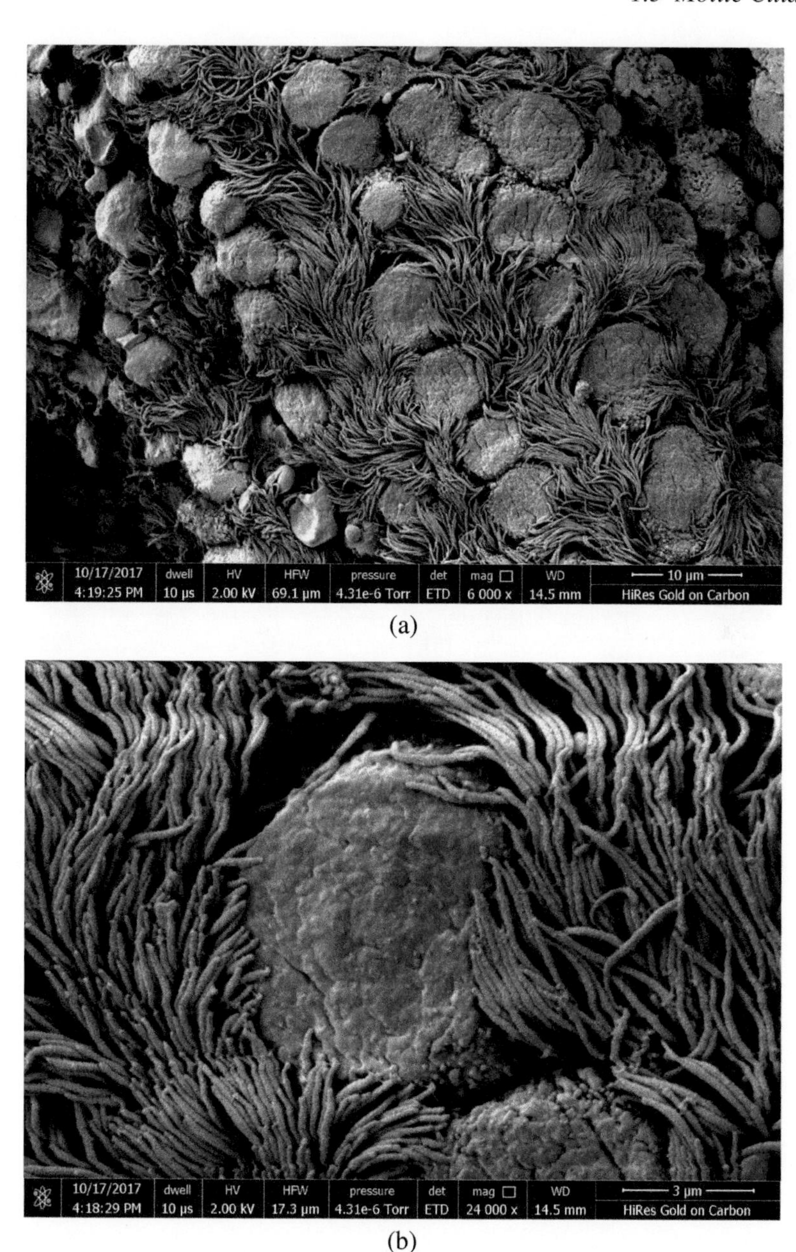

Figure 1.1 Scanning electron micrographs of rat trachea segments at two magnifications. Cilia sprouting from ciliated epithelial cells can be seen interspersing mucous-secreting goblet cells. Preparation method was glutaraldehyde fixation followed by HMDS dehydration. Copyright Samuel Powell 2018.

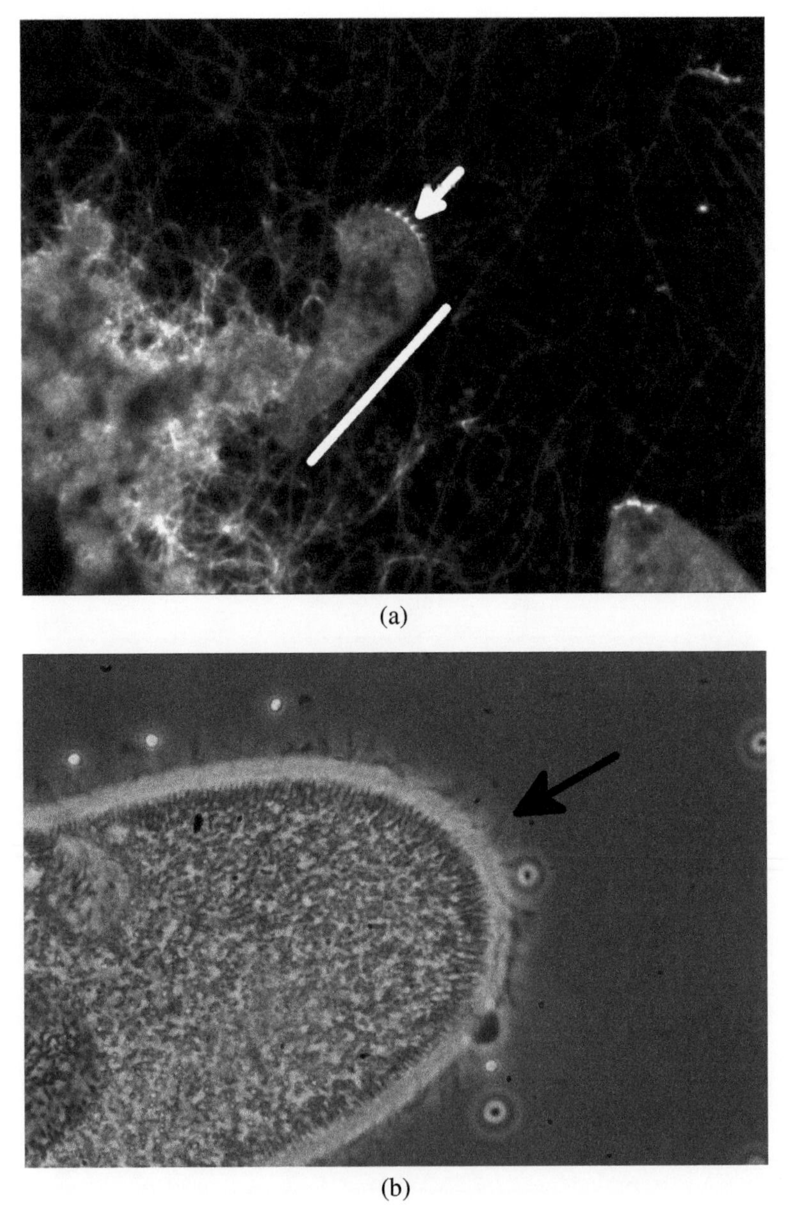

(a)

(b)

Figure 1.2 Light micrographs showing two varieties of ciliate. (a) *Stentor coeruleus*, dark-field illumination, original magnification ×40. The organism (indicated by a bar) is anchored to organic detritus. Cilia are arrowed. (b) Anterior tip of *Paramecium caudatum*, phase contrast, original magnification ×400. Cilia are arrowed; for scale, cilia are 10μm long. Copyright Richard Mayne 2018.

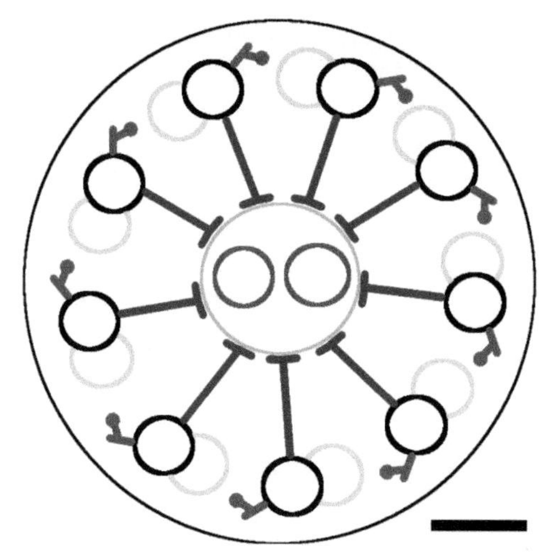

Figure 1.3 Diagram showing a motile cilium in cross section. Around the periphery lie the nine microtubule doublets (black A subunits and gray B subunits). Dynein arms (red) emanate from A tubules and their active sites sit close to the B subunit doublets they interact with in the presence of ATP. The doublets are articulated onto the sheath (green) that surrounds the central microtubule pair (pink) via radial spikes (blue). Scale bar (lower left) is approximately 50 nm; the entire cilium is approximately 300 nm in diameter (dynein arms, radial spokes, and sheath not to scale). Copyright Richard Mayne 2018.

1.3.2 The Ciliary Beat

The ciliary beat is formed from two components (Figure 1.4). First, the "active" or "power" stroke, where the cilium pivots about its base and remains straight along its longitudinal axis, is the component of the beating that generates motive force. The second component is the "return" or "passive" stroke, where the cilium bends sequentially base-to-tip about its longitudinal axis in the opposite direction to the power stroke. The return stroke's pattern generates very little motive force, thereby making the beating cycle a highly effective pattern of motion for providing net movement in a single direction. In the horizontal axis, the beating cycle follows a slightly elliptical pattern [17].

Motile cilia beat continuously under normal conditions, but beating cycles may be modified adaptively with regard to the beating frequency and orientation. The mechanisms that influence these modifications are differed between cell types/species, but they typically result from modulation of intracellular calcium ion concentration ([Ca^{2+}]) resulting from stimulus-induced membrane depolarization. For example, in *Paramecium* spp., a head-on collision

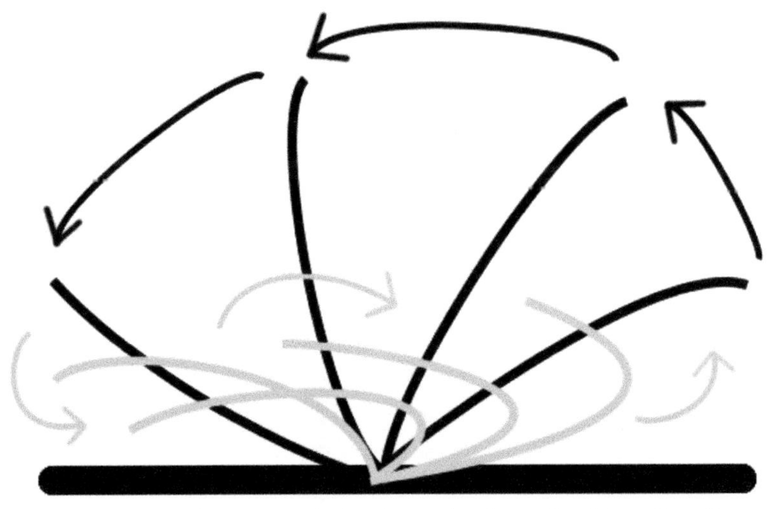

Figure 1.4 Diagram to demonstrate the beating cycle of a motile cilium. The cilium is rigid during the power stroke (black), before bending base-to-tip on its return during the rest stroke (gray). Adapted from [24].

with a solid object (i.e., tactile stimulus) will trigger a transient reversal in ciliary beat direction, causing the organism to "reverse" [20].

From a biophysical perspective, inertial forces are almost non-existent in the immediate environment of the motile cilium, which means that local fluid movements are akin to those observed in low Reynolds number environments [21]. It is thought that this phenomenon underlies one of the most fascinating aspects of the motile cilia arises: the emergent property of metachronism. A metachronal wave is an asynchronous traveling wave across the surface of a multiciliated cell (Figure 1.5); in other words, in a two-dimensional array $n \times m$, cilia will beat at the same frequency but a different phase to the axis of the effective stroke, n, and synchronously with cilia along the m-axis, i.e., perpendicularly to n [22]. Metachronal waves exist to produce smooth patterns of motion: consider how cilia beating in unison would produce a jerky motion akin to a boat powered by oars.

Despite the mechanisms of ciliary beating being reasonably well characterized, it is poorly understood how these emergent patterns occur when no centralized cellular "wave controller" exists. By extension, this implies that only local interactions are involved in metachronal wave formation; it is from this observation that the widely accepted hypothesis that hydrodynamic

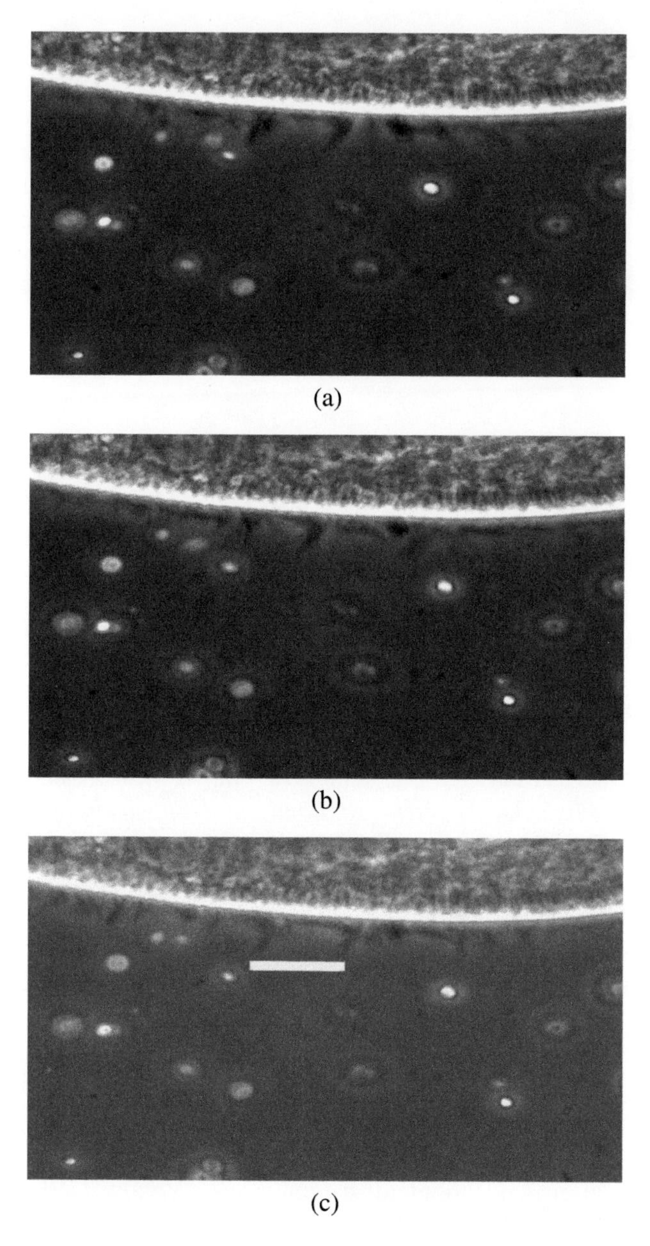

Figure 1.5 Video micrographs showing ciliary beating along the dorsal surface of the ciliate *Paramecium caudatum*. These sequential images, separated by 5-ms intervals, show several hundred cilia beating at approximately 30 Hz. A metachronal wave is marked in with a white bar; for scale, cilia are 10μm long. Original magnification $\times$100. Copyright Richard Mayne 2018.

coupling between neighboring cilia underlies metachronal wave formation was formed [17, 19, 21]. It must be emphasized however, that the vast majority of the evidence in favor of this theory is numerical and modeling studies. It remains, furthermore, unclear whether the cytoskeletal components of ciliated cells (predominantly microtubules and microfilaments), which are anchored to the basal bodies of cilia in various patterns, participate in metachronal wave propagation [23]. It is therefore important to remember that motile cilia are not simply actuators, but independent sensors with only local (neighbor-to-neighbor) communication.

References

[1] van Leewenhoeck A. Observations, Communicated to the Publisher by Mr. Antony van Leewenhoeck, in a Dutch Letter of the 9th of October. 1676. Here English'd: Concerning Little Animals by Him Observed in Rain-Well-Sea and Snow Water; As Also in Water Wherein Pepper Had Lain. In: *Philosophical Transactions of the Royal Society of London*, 12(133-142):821–831, 1677.

[2] Müller O. Animalcula infusoria; fluvia tilia et marina, que detexit, system-atice descripsit et ad vivum delineari curavit. N. Molleri, London, 1786 [In Latin].

[3] Zimmerman K. Beiträge zur kenntniss einiger drušen und epithelien. *Archiv für Mikroskopische Anatomie und Entwicklungsgeschicte*, 52(3):552–706, 1898 [In German].

[4] Sorokin S. Reconstructions of the centriole formation and ciliogenesis in mammalian lungs. *Journal of Cell Science*, 3:207–230, 1968.

[5] Singla V, and Reiter J. The primary cilium as the cell's antenna: signaling at a sensory organelle. *Science*, 313(5787):629–633, 2006.

[6] Taschner M, and Lorentzen E. The intraflagellar transport machinery. *Cold Spring Harbor Perspectives in Biology*, 8(10):a028092, 2016.

[7] Szymanska K, and Johnson C. The transition zone: an essential functional compartment of cilia. *Cilia*, 1(1):10, 2012.

[8] Nigg E, and Stearns T. The centrosome cycle: Centriole biogenesis, duplication and inherent asymmetries. *Nature Cell Biology*, 13(10):1154–1160, 2011.

[9] Seeley E, and Nachury M. The perennial organelle: assembly and disassembly of the primary cilium. *Journal of Cell Science*, 123(4):511–518, 2010.

[10] Huangfu, D., Liu, A., Rakeman, A. S., Murcia, N. S., Niswander, L., and Anderson, K. V. Hedgehog signalling in the mouse requires intraflagellar transport proteins. *Nature* 426:83–87, 2003. doi: 10.1038/nature02061
[11] Wheway G, Nazlamova L, and Hancock J. Signaling through the primary cilium. *Frontiers in Cell and Developmental Biology*, 8(6):8, 2018.
[12] Wheway G, Schmidts M, Mans DA, et al. An siRNA-based functional genomics screen for the identification of regulators of ciliogenesis and ciliopathy genes. *Nature Cell Biology*, 17(8):1074–1087, 2015.
[13] Boldt K, Van Reeuwijk J, Lu Q, et al. An organelle-specific protein landscape identifies novel diseases and molecular mechanisms. *Nature Communications*, 7:1–13, 2016.
[14] Waters A, and Beales P. Ciliopathies: an expanding disease spectrum. *Pediatric nephrology*, 26(7):1039–1056, 2011.
[15] Hassounah N, Bunch T, and McDermott K. Molecular pathways: the role of primary cilia in cancer progression and therapeutics with a focus on hedgehog signaling. *Clinical Cancer Research*, 18(9):2429–2435, 2012.
[16] Adl SM, Leander BS, Simpson AGB, et al. Diversity, nomenclature, and taxonomy of protists. *Systematic Biology*, 56(4):684–689, 2007.
[17] Lindemann CB, and Lesich KA. Flagellar and ciliary beating: the proven and the possible. *Journal of Cell Science*, 123(4):519–528, 2010.
[18] Pigino G, and Ishikawa T. Axonemal radial spikes. *Bioarchitecture*, 2(2):50–58, 2012.
[19] Vilfan A, and Julicher F. Hydrodynamic flow patterns and synchronization of beating cilia. *Physical Review Letters*, 96(5):1–4, 2006.
[20] Tamm S. Ca^{2+} channels and signalling in the cilia and flagella. *Trends in Cell Biology*, 4(9):305–310, 1994.
[21] Golestanian R, Yeomans JM, and Uchida N. Hydrodynamic synchronization at low Reynolds number. *Soft Matter*, 7(7):3074, 2011.
[22] Brooks E and Wallingford J. Multiciliated cells: a review. *Current Biology*, 24(19):R973–R982, 2014.
[23] Werner M, Hwang P, Huisman F, et al. Actin and microtubules drive differential aspects of planar cell polarity in multi-ciliated cells. *Journal of Cell Biology*, 195:19–26, 2011.
[24] Whiting JGH, Mayne R, and Adamatzky A. A Parallel Modular Biomimetic Cilia Sorting Platform. *Biomimetics*, 3(5):1–15, 2018.

2

Genetic Engineering of Ciliated Cells

Gabrielle Wheway

Centre for Research in Biosciences, Department of Applied Sciences,
Faculty of Health and Applied Sciences, University of the West of England,
Bristol BS16 1QY, United Kingdom
E-mail: Gabrielle.Wheway@uwe.ac.uk

The cilium is a discrete organelle and an excellent model for systems biology studies, being less complex than a complete cell, tissue, organ, or organism. Owing to this fact, and the fact that mutations in genes encoding ciliary proteins lead to medically significant conditions, cilium genetics has been the focus of intensive research over the past 15 years. As a result, the genes required for making and maintaining a functioning cilium are relatively well characterized, although our understanding of these processes is far from complete.

With this knowledge, it is comparatively easy to manipulate cilium function through genetic engineering; we understand which genes will affect different parts of cilium function to produce particular phenotypes. The ability to engineer such defects provides us with resources to gain insight into cilium structure and function, and move toward the development of therapeutics to treat cellular defects underlying ciliopathies, as well as manipulate cilia for biocomputation purposes.

The majority of mammalian adherent epithelial cell lines can be induced to grow primary cilia in cell culture conditions, through simple serum starvation. Recent advances in stem cell culturing techniques now make it possible to culture cells which can be differentiated to form motile cilia. Many of these ciliated cell lines are easily genetically manipulated using an ever-expanding array of molecular tools from short interfering RNA (siRNA) to Clustered Regularly Interspersed Short Palindromic Repeats (CRISPR)/Cas9.

This chapter provides an introduction to these methods and serves as a guide to the current state of the art in genetic engineering of ciliated cells.

2.1 Introduction

Functional genomics studies, in which the function of every gene in the genome is individually tested for its contribution to a particular phenotype, such as ciliogenesis, suggest that around 10–15% of the mammalian genome is required for ciliogenesis and cilium function [1, 2].

Comparative genomics studies of ciliated organisms from unicellular ciliates to man show that many of these genes are common to all ciliated organisms, and highly conserved through evolution, with some species-specific differences (for example, the vast expansion in number of olfactory receptors in the cilium membrane of mice). Similarly, comparative proteomics of ciliated organisms has allowed identification of core cilia proteins common across different ciliated species.

Broad genomics studies provide researchers with useful resources for studying cilia. GenomeRNAi (http://www.genomernai.org/) [3] is a publicly accessible online resource with simply search facility for interrogating functional genomics datasets. A similar online resource, Cildb (http://cildb.cgm.cnrs-gif.fr/) [4, 5], is a database of protein expression in cilia of a diverse range of organisms.

For the convenience of the scientific community, a comprehensive comparison of functional genomics, comparative genomics, and comparative proteomics datasets has produced the "Cilia Gold Standard" list of *bona fide* cilia genes and proteins [6]. Such a list provides high-confidence positive and negative control genes which will have specific effects on cilia when manipulated, a useful resource for any investigator wishing to model and investigate cilia form and function.

Using this information, investigators can select candidate genes for manipulation to produce model cell systems for the study of cilium biology.

This chapter will discuss selection and growth of appropriate cells, transfection of genetic material into the cells, and application of the following genetic engineering technology:

- RNA interference using short interfering RNA (siRNA) and short hairpin RNA (shRNA)
- Fluorescent gene tagging and its variants such as split GFP
- Optogenetics

- BioID
- TAP tagging
- Genome editing using transcription activator-like effector nucleases (TALENs), Zn finger nucleases, CRISPR/Cas9, and base editing.

2.2 Cell Selection and Growth

When selecting a cell type for *in vitro* studies, a number of factors must be considered:

- Relevance of the origin of the cell type to the research question. This includes the species of origin, and the tissue of origin.
- Transfection potential of cells.
- Ease with which these cells can be genetically manipulated. This includes not only a consideration of technical difficulty using particular cells but also the resources available to achieve this, including whether the genome of the species of origin has been fully sequenced and characterized.
- Ease with which the cells can be grown.
- Adherence of the cells to growth surface in culture. Adherence of cells to a culture surface immobilizes the cells to enable imaging and other assays.
- Availability and cost.

2.2.1 Cells with Primary Cilia

While most cells have some capacity for growing primary cilia in culture, the most well-characterized commercially available ciliated cell lines are summarized in Table 2.1, all of which are available from American Type Culture Collection (ATCC; https://www.lgcstandards-atcc.org/). These are all fully adherent, immortalized cell lines, with almost unlimited growth potential in *in vitro* culture.

Cilia growth is tightly linked to the cell cycle, as the basal body which nucleates the ciliary axoneme is derived from the centriole, the structure which organizes the spindle during cell division. Cilia grow on cells which are in G0 phase, having exited the cell cycle, or during G1/S phase of cycling cells. To induce exit from the cell cycle and hence induce ciliogenesis, the removal of serum from the culture medium is sufficient to induce ciliogenesis in most of these cell types. This process of "serum starvation" is a simple and easy way to induce ciliation of cells. mIMCD3 cells are a particularly

Table 2.1 The most commonly used commercially available primary ciliated cell lines, their species and tissue of origin, rate at which they grow cilia *in vitro*, ease with which they can be transfected, and ATCC reference number

Cell Line Name	Species Origin	Tissue Origin	Ciliation *in vitro*	Transfection	ATCC Ref
A6	*Xenopus laevis*, frog, South African clawed	Kidney	**	?	CCL-102
ARPE-19	*Homo sapiens*	Retina – pigmented epithelium	**	*	CRL-2302
HEK293(T)	*Homo sapiens*	Embryonic kidney	*?	****	CRL-1573; CRL-6216
HeLa	*Homo sapiens*	Cervix – adeno-carcinoma	*	***	CCL-2
hTERT-RPE1	*Homo sapiens*	Retina – pigmented epithelium	**	**	CRL-4000
LLC-PK1	*Sus scrofa* (pig)	Kidney – proximal tubule	**	?	CL-101
mIMCD-3	*Mus musculus* (mouse)	Kidney – inner medullary collecting duct	***	***	CRL-2123
MDCK	*Canis familiaris* (dog)	Kidney – distal tubule/ collecting duct	**	**	CCL-34
NIH/3T3	*Mus musculus* (mouse)	Fibroblast	***	***	CRL-1658

useful cell line for studying cilia, as they grow many, long cilia even without serum starvation, during the G1/S phase of the cell cycle (Figure 2.1). The cilia of this cell type also tend to be longer (5–15 microns in length rather than 0–5 microns) than many of the other cell lines, although A6 cells have been reported to grow very long cilia up to 50 microns long.

Ciliogenesis is also tightly linked to cell polarity. Cilia form on the apical cell surface of the polarized cells, after the basal body, derived from the mother centriole, migrates to the apical cell surface. Some cells polarize more easily in culture than others. Some, including A6 cells and MDCK cells, require growth on porous filters to allow the underside of the cells to contact the growth media to allow proper polarization and ciliogenesis [7, 8]. While HEK293T cells are not easily ciliated, they can be induced to form cilia using this growth method and extended serum starvation [9] (Figure 2.3). HeLa

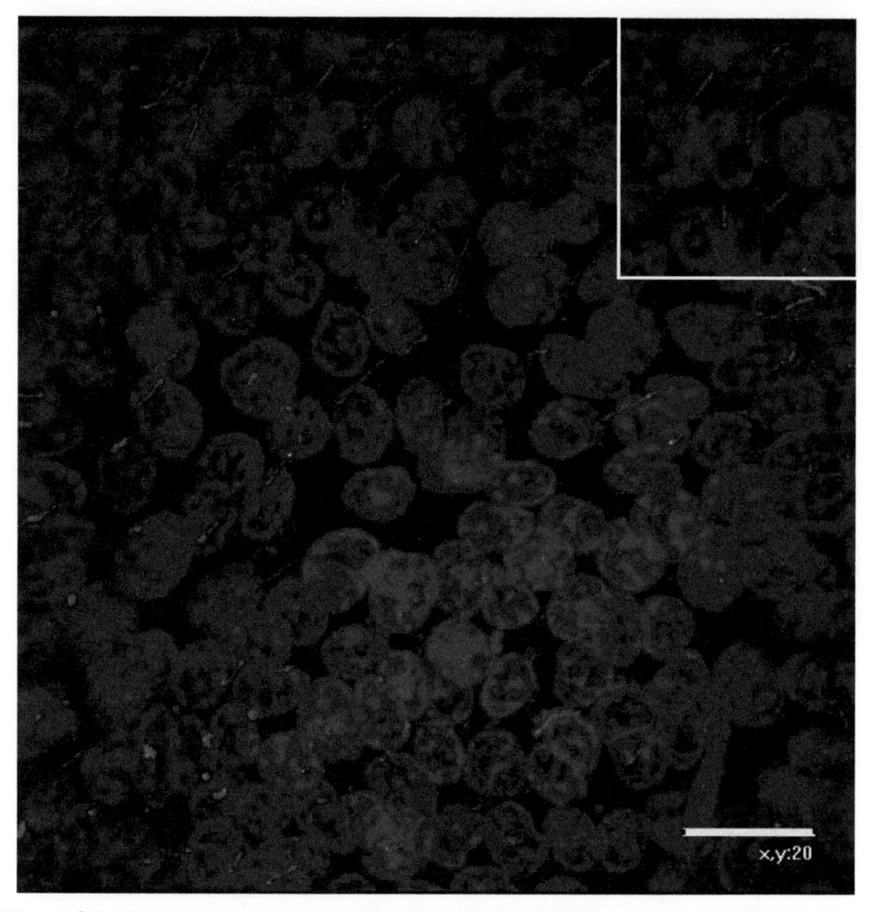

Figure 2.1 Mouse Inner medullary collecting duct (mIMCD3) adherent immortalized cell line with primary cilia stained in red (anti-acetylated alpha tubulin) and basal body in green (anti-IFT80). Scale bar = 20 μm. Copyright G. Wheway.

cells are similarly not a typical ciliated cell model, but are included here due to recent evidence that they robustly grow cilia *in vitro* [10].

It can be noted that the availability of well-characterized adherent, easily polarized, and ciliated cell lines is limited. As an alternative to immortalized cell lines, investigators may consider using primary cells, which have been derived directly from a living organism, such as fibroblasts from a human skin biopsy or mouse embryo (Figure 2.2) [11, 12]. These provide cells from a specific genetic background, such as a patient with a genetic disease, or transgenic mouse, but are difficult to transfect, and will only grow in culture

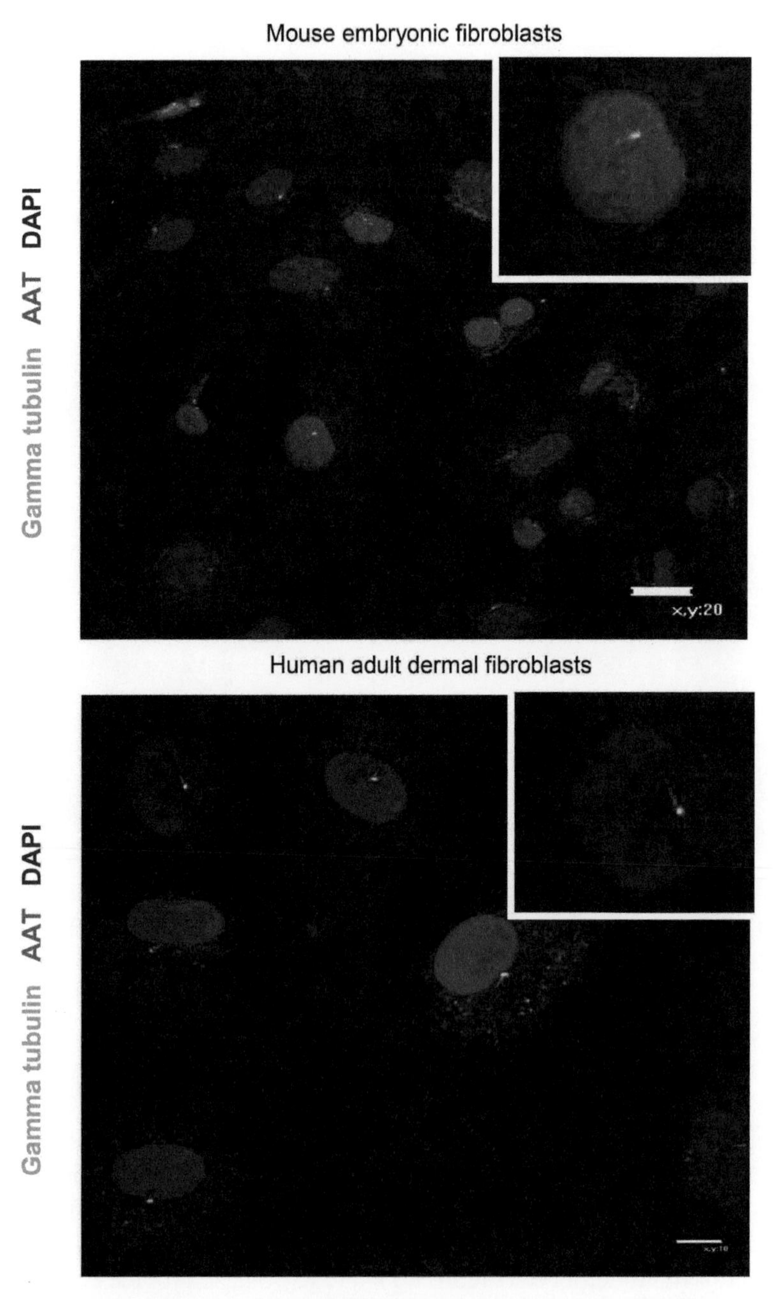

Figure 2.2 Mouse embryonic fibroblasts (top) and adult human dermal fibroblasts (bottom) with primary cilia stained in red (anti-acetylated alpha tubulin (AAT)) and basal body in green (anti-gamma tubulin). Scale bar = 20 μm (top) and 10 m (bottom). Copyright G. Wheway.

for approximately seven passages (the process of diluting cells into fresh culture plates and media to allow continued growth).

2.2.2 Cells with Motile Cilia

While there are no adherent immortalized cell lines which can grow motile cilia, motile ciliated cells can be grown from induced pluripotent stem cells and, theoretically, from embryonic stem cells (Figure 2.3).

This is achieved through providing culture conditions which simulate the *in vivo* environment of stem cells developing toward a tracheal epithelial cell fate. The culture process involves stimulation with Wnt3a and activin in the first 3 days to form definitive endoderm, followed by inhibition of activin/TGF-β with SB431542 and inhibition of BMP4 with noggin on days 4–9 to form ventral anterior foregut. The cells are then exposed to FGFs, and at day 17 transferred to an environment with an air–liquid interface, and stimulated to form multiciliated cells through notch inhibition via DAPT treatment. After 45 days, a mature tracheal epithelium is formed, with multiciliated cells [13].

Study of motile ciliated cells is usually conducted by studying cells *in vivo*, or on *ex vivo* cultures such as mouse tracheal explants or nasal brush samples from patients with primary ciliary dyskinesia. Much characterization of motile cilia has been achieved by single-celled flagellated organisms *Trypanosoma brucei* [14] and *Chlamydomonas reinhardtii* [15] which possess one and two flagella, respectively, and *Paramecium* which possesses around 5000 flagella per cell. These model systems offer numerous advantages, including ease of culture and biochemical purification of motile cilia and ease of genetic manipulation. However, these single-celled organisms are free-living, using their flagella for motility, and are difficult to culture in adherent sheets, significantly reducing their utility for bioengineering and biocomputation. For these reasons, these cells will not be dealt with in this chapter. For a useful review of cilia model organisms, see reference [16].

2.3 Transfection of Cells

In order to genetically manipulate a cell, short linear or circular DNA or RNA fragments must be inserted into the cells, from which exogenous genes can be expressed, or with which endogenous genes or transcripts can be modified. This is typically achieved using lipid-based transfection reagents which coat the nucleic acid with a cationic lipid to allow it to pass across

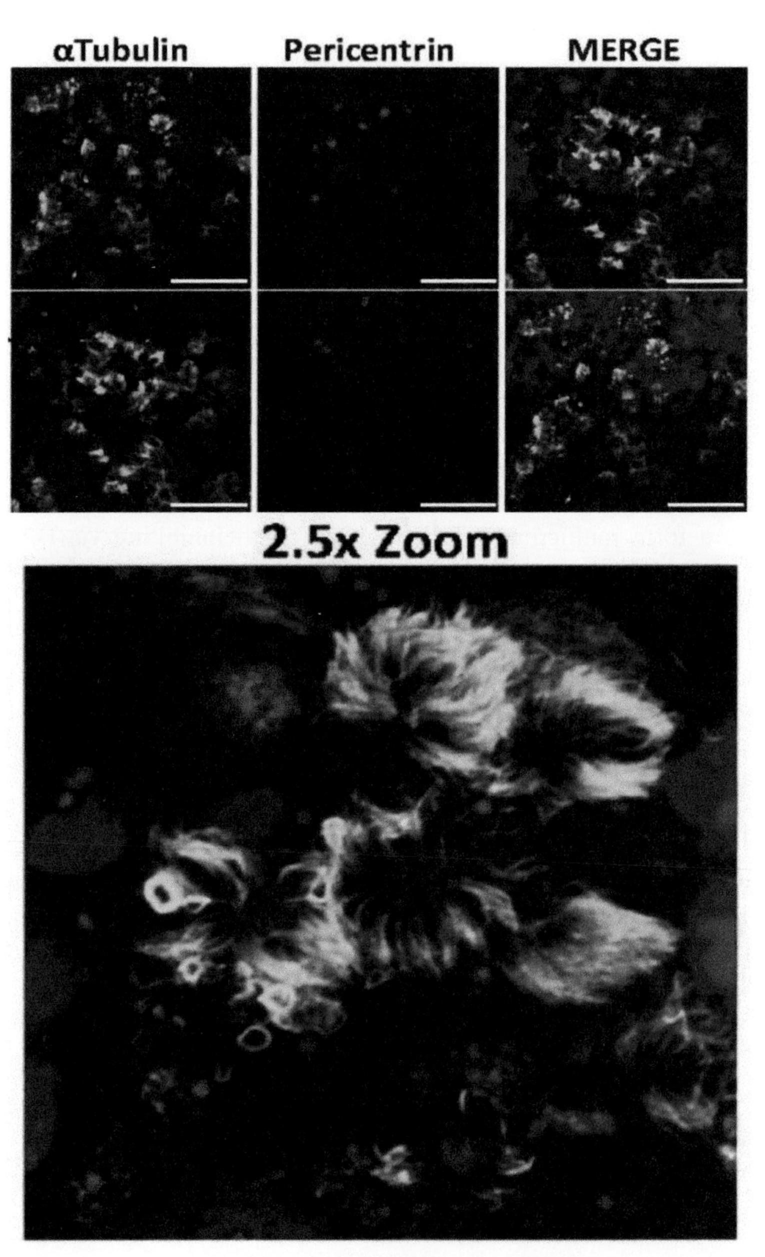

Figure 2.3 Tracheal epithelial-like multiciliated cells derived from human induced pluripotent stem cells (iPSCs). Cilia axoneme stained in cyan (anti-acetylate alpha tubulin) and basal bodies in red (anti-pericentrin). Scale bar = 100 μm. Modified from reference [13].

the phospholipid cell membrane. This is a straightforward method, and can be achieved with commercial proprietary reagents such as Lipofectamine^{TM} (Life Technologies), or with inexpensive polyethylenimine. The efficiency with which different cell lines of different origins can be transfected using this method varies greatly, and must be considered when selecting an appropriate cell line for study, as the efficiency of delivery of nucleic acids to host cells can be a significant limiting factor in experiments. It is as an excellent transfection host that the HEK293T cell line is so widely used. However, these cells do not readily form cilia. mIMCD3 cells are a good choice for cilia studies, as they readily grow cilia and can be transfected efficiently, but are of mouse origin. While the human hTERT-RPE1 cell line is a good choice of cell line for growing cilia, it is less readily transfected.

Primary cells, stem cells, and some cell lines are very difficult to transfect. Alternative methods of nucleic acid delivery include electroporation, nucleofection, and delivery by retrovirus or lentivirus. These methods are more technically challenging, and viral methods must be conducted in a biosafety level II facility, whereas generally other eukaryotic cell culture techniques are biosafety level I.

2.4 RNA Interference

RNA interference involves transfection of cells with an siRNA molecule or plasmid encoding an shRNA molecule complementary in sequence to the mRNA sequence of a target gene. Binding of siRNA or shRNA to the mRNA produces double-stranded RNA in the cell, which is rapidly degraded by cellular machinery. This results in reduction of protein production, effectively knocking down levels of a specific protein in cells. siRNA targeting of cilia genes can produce striking cilia phenotypes (Figure 2.4), and libraries of siRNAs targeting all genes in the genome can now be bought commercially, allowing screens of the entire genome for genes regulating cilia (Figure 2.5) [2]. shRNA knockdown can have a more stable effect on protein reduction, if the plasmid is stably incorporated into the host cell (Figure 2.6).

2.5 Fluorescent Protein Labeling

Green fluorescent protein, first extracted from *Aequorea victoria* jellyfish [17], and its derivatives which fluoresce at different wavelengths, is a small protein (27 kDa) which can be used as a visual tag to track protein

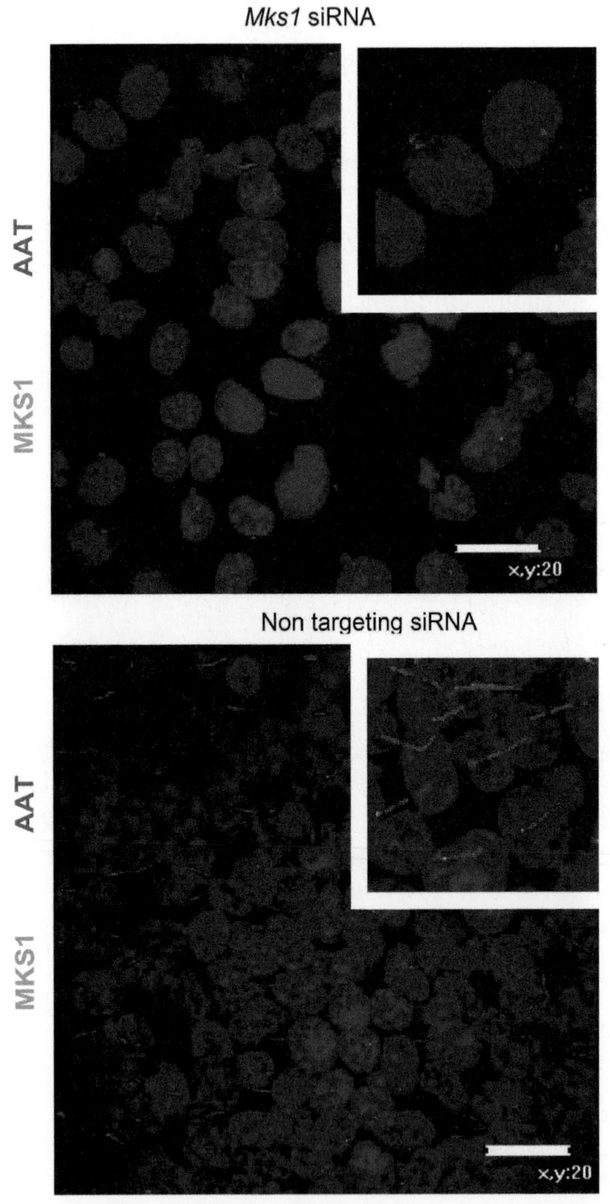

Figure 2.4 Loss of cilium axoneme in mouse IMCD3 cells after siRNA knockdown of *Mks1*, a basal body protein (top), compared to cilia in cells transfected with a non-targeting siRNA with scrambled sequence (bottom). Cilia axoneme is labeled in red with acetylated alpha tubulin (AAT), MKS1 is labeled in green, and nuclei are stained blue with DAPI. Copyright G. Wheway.

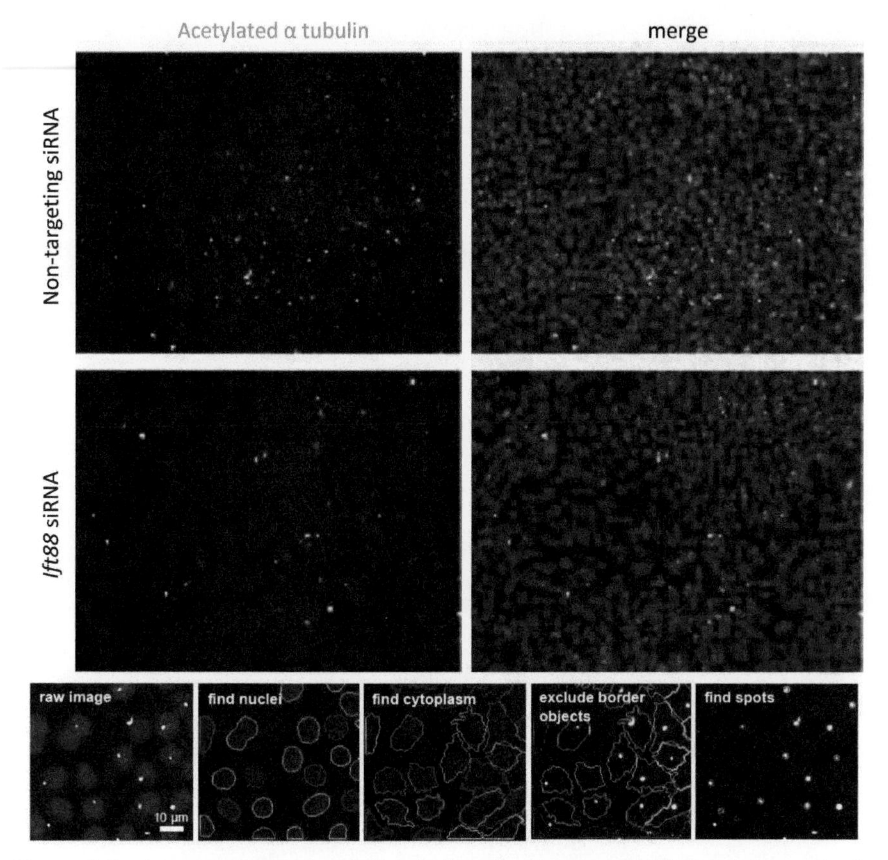

Figure 2.5 Automated image analysis of ciliated cells after siRNA knockdown, for the purposes of high-throughput screening. Modified from reference [2].

localization, through cloning of this gene into a plasmid which expresses a gene of interest. The resultant protein will have a fluorescent tag, to allow imaging of protein localization, in fixed cells (Figure 2.7) and in live cells (Figure 2.8).

A recent advance in this technology, so-called "split GFP," involves tagging the gene encoding a protein of interest with a segment of DNA encoding only the 11th β-strand of super-folder GFP (GFP11) or mCherry (mCherry11) [18]. This offers many advantages to conventional fluorescent tagging. At the genetic level, the further reduced size of the DNA fragment requiring cloning significantly enhances the ability to "knock-in" this tag into genomic DNA using CRISPR/Cas9 or other genome editing techniques

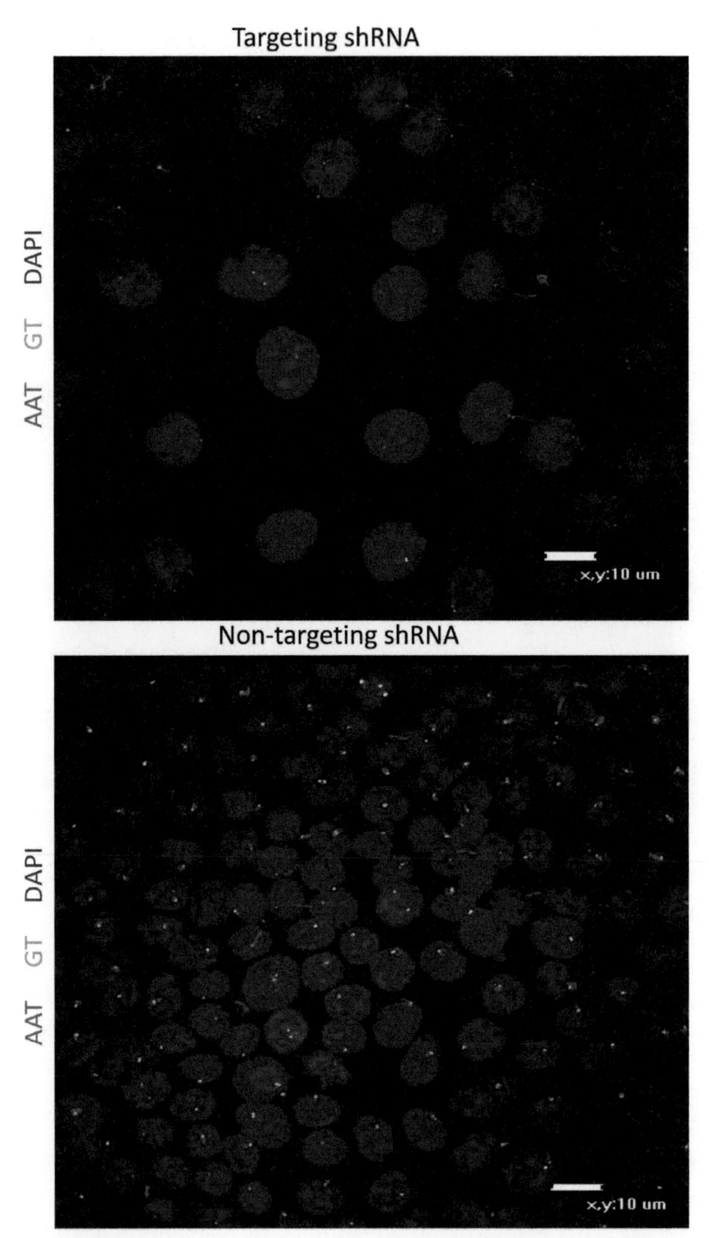

Figure 2.6 Lengthening of cilium axoneme in mouse IMCD3 cells after shRNA knockdown of a gene regulating cilium length (top), compared to cilia in cells transfected with a non-targeting siRNA with scrambled sequence (bottom). Cilia axoneme is labeled in red with acetylated alpha tubulin (AAT), gamma tubulin is labeled in green, and nuclei are stained blue with DAPI. Scale bars = 10 μm. Copyright G. Wheway.

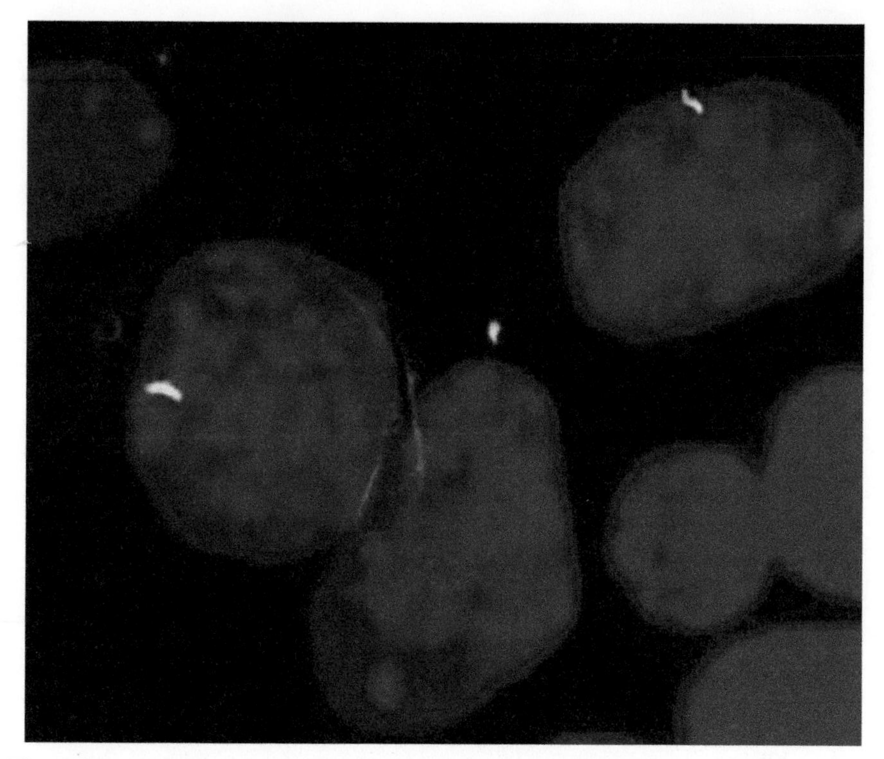

Figure 2.7 Lebercilin, encoded by LCA5, tagged with enhanced yellow fluorescent protein (eYFP) in IMCD3 cells, fixed, and stained with DAPI and acetylated alpha tubulin (red). Copyright G. Wheway.

(see later). At the protein level, the smaller tag size reduces the effect of the protein tag on endogenous protein function and localization, for example, reducing the likelihood, it will be retained in the ER. Tandem arrays of this 11th β-strand can be used to enhance the stability of the fluorescent signal (Figure 2.9) [19].

2.6 Optogenetics

Optogenetics is a system which uses light-responsive ion channels (channelrhodopsins), and light-responsive ion pumps (halorhodopsins and archaerhodopsins) to produce specific, rapid, targeted cellular responses [20]. As cAMP is known to play an important role in regulating ciliary beat frequency, cytogenetics has been applied to controlling sperm flagellum dynamics via

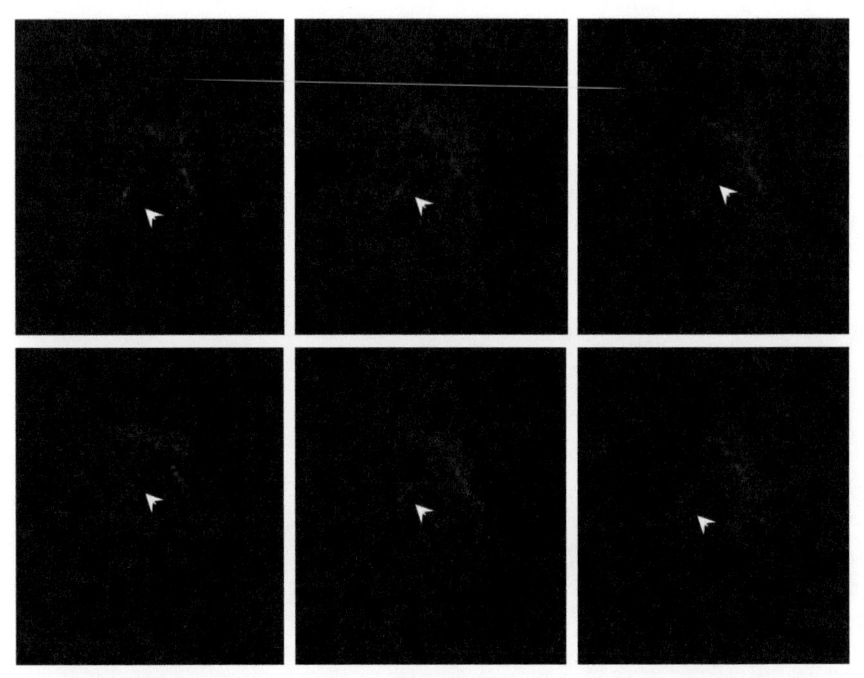

Figure 2.8 Stills from a live cell image capture of lebercilin-eYFP in live IMCD3 cells. Copyright G. Wheway.

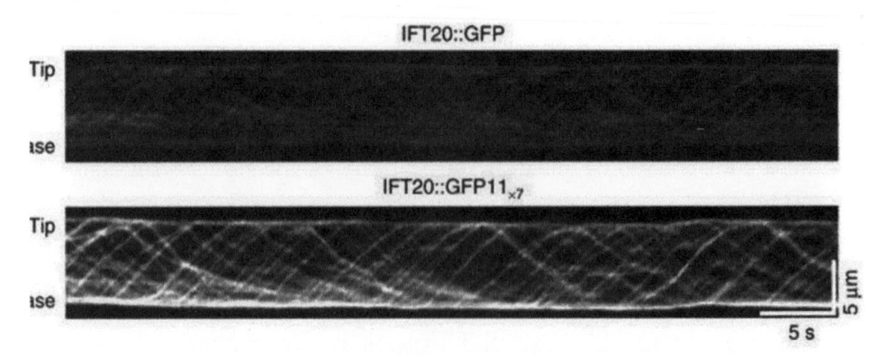

Figure 2.9 Kymographs showing anterograde and retrograde transport of IFT20 tagged with GFP (IFT20::GFP, top) and a tandem array of seven-"split" GPF11 β-strands (IFT20::GFP11 $\times$ 7) in IMCD3 cells. Modified from reference [19]. Reproduced with kind permission from Prof. Bo Huang.

light activation of cAMP in flagella (Figure 2.10) [21], using photoactivated adenylyl cyclase bPAC [22, 23].

Optogenetics represents an exciting opportunity for cilia bioengineering and biocomputation through the accurate and specific activation of cilia using light. The use of laser light of discrete wavelengths, focused on specific focal points, can provide literally pinpoint accuracy of manipulation of cells, with high spatial and temporal resolution. Optogenetics has had huge impacts on neuroscience, allowing elucidation of functions of individual neurons. This includes ciliated neurons such as the Kolmer–Agduhr cell, which possess motile cilia for the movement of cerebrospinal fluid [24].

2.7 Proximity-Dependent Biotinylation (BioID)

Proximity-dependent biotinylation (BioID) [25] involves expressing bait proteins tagged with biotin ligase from an exogenous plasmid in a cell type of interest. Biotin ligase catalyzes biotinylation of proteins in proximity to the bait protein, and biotinylated proteins are subsequently purified from cell extract by virtue of biotin's high affinity for streptavidin, and identified using tandem mass spectrometry. This approach has been used to successfully elucidate protein–protein interactions at the base of the cilium where the basal body becomes the axoneme, through biotin ligase tagging of proteins known to localize to distinct subcompartments of the cilium/basal body interface: centriolar satellites, centrioles, appendages, and transition zone (Figure 2.11) [26].

2.8 Tandem Affinity Purification (TAP)

Tandem affinity purification (TAP) involves expressing bait proteins tagged with FLAG octapeptide and strep-tag II from an exogenous plasmid in a cell type of interest. Bait proteins and their interactants (both direct and indirect) are subsequently purified from cell extract using FLAG antibody and strep beads, and proteins are identified using tandem mass spectrometry [27]. This system can purify whole complexes intact, from which proteins bound with different affinities can be serially eluted using differing SDS concentrations [28]. This can also be developed into a quantitative technique via the use of cell growth media with different isotopes of particular amino acids, using an approach called stable isotope labeling in cell culture [29]. This system has further elucidated the interactome of the primary cilium (Figure 2.12) [30].

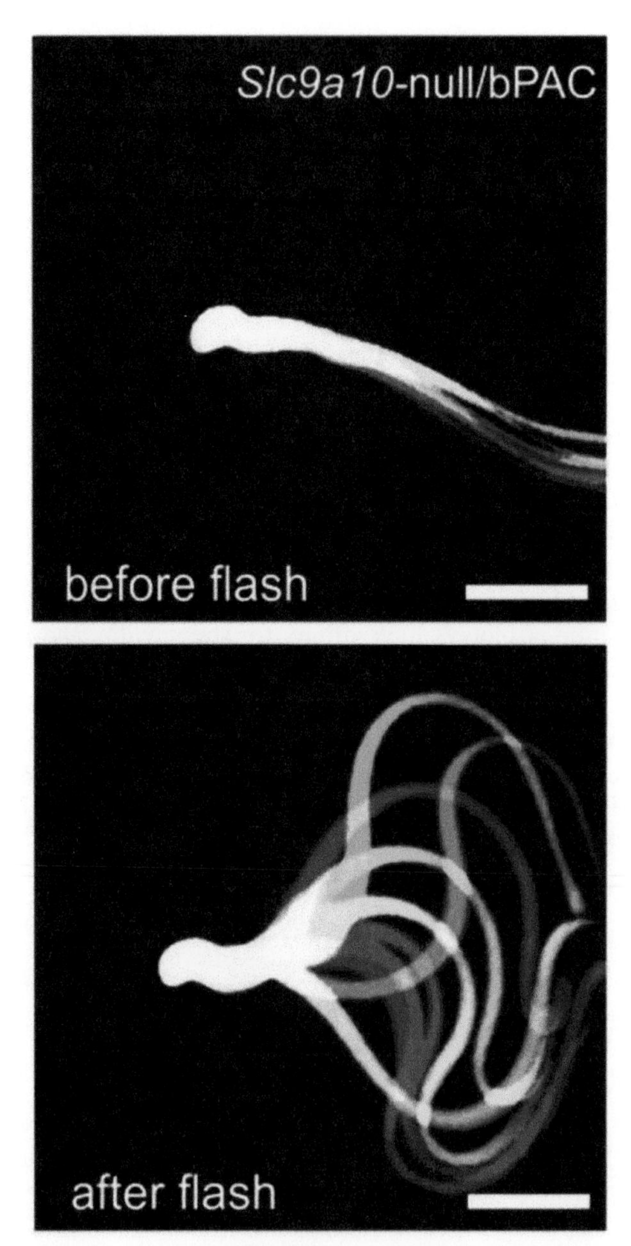

Figure 2.10 Flagellar waveform of sperm from infertile mice genetically modified to express photoactivated adenylyl cyclase. The image on the left shows normal flagellar beating and the image on the right shows and flagellar beating after cAMP activation via UV stimulation. Modified from reference [21]. Reproduced with kind permission of Prof. Dagmar Wachten.

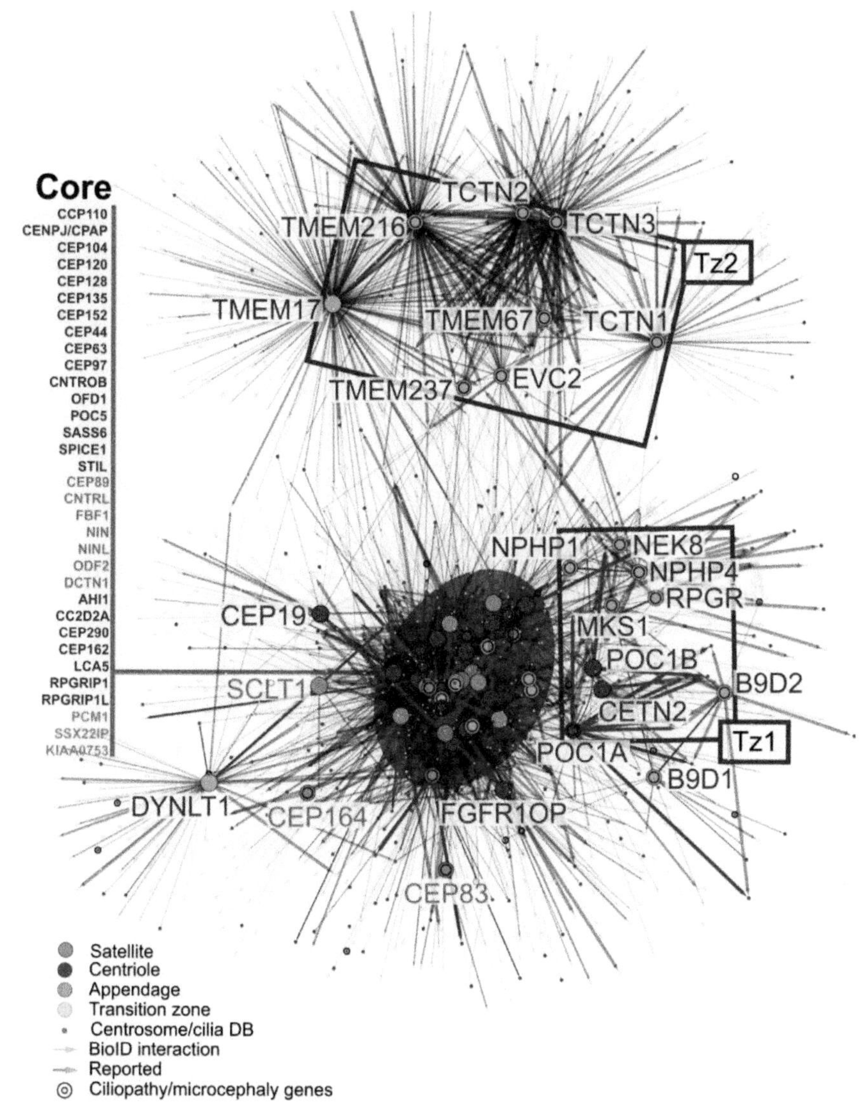

Figure 2.11 Topology map (based on peptide count sum of two mass spectrometry runs) of the basal body-cilium BioID interactome. Modified from reference [26]. Reproduced with kind permission of Prof. Laurence Pelletier and Prof. Brian Raught.

2.9 Genome Editing Using TALENs, Zn Finger Nucleases, CRISPR/Cas9, and Base Editing

Over the past 20 years, there has been steady development of the use of engineered nucleases, enzymes which bind and digest specific regions of

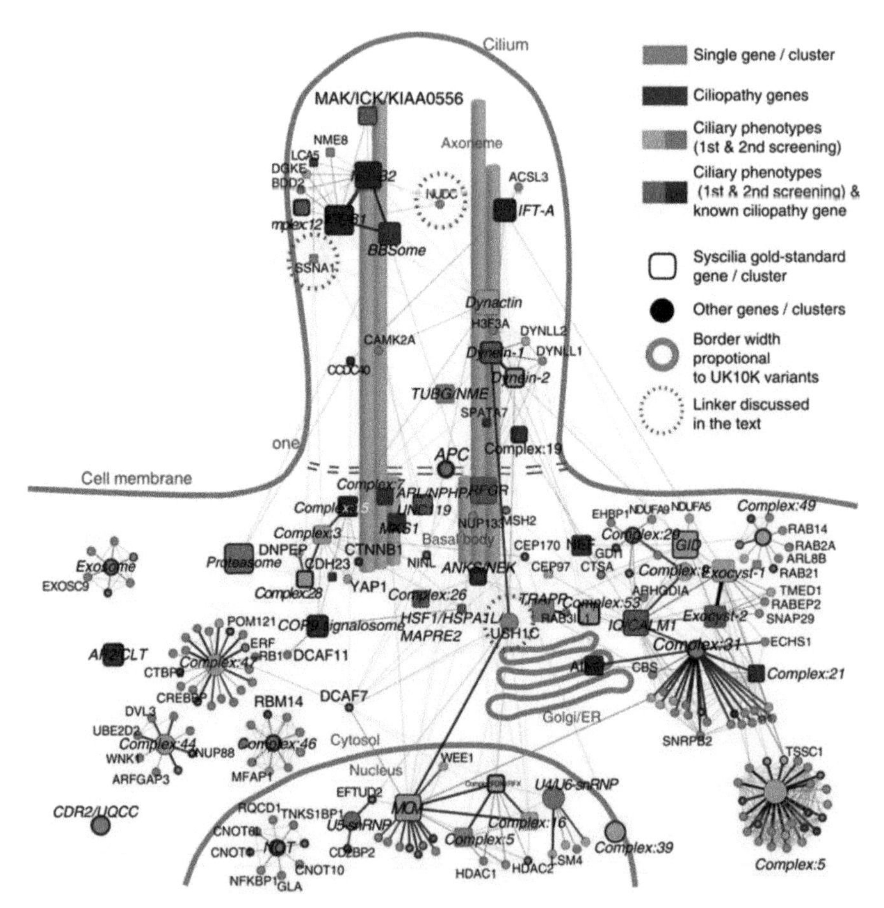

Figure 2.12 The cilium interactome, as identified by TAP. Modified from reference [30]. Reproduced with kind permission of Dr. Karsten Boldt and Prof. Ronald Roepman.

DNA, to make changes to DNA sequences in the host genome of target cells. Nucleases introduce double-strand breaks (DSBs) into DNAs, which are inaccurately repaired by a process called non-homologous end joining. This often leads to insertions or deletions in DNA, which can knock out the function of a gene. At a lower rate of efficiency, DSBs can be repaired by homology-directed repair, which repairs the DNA using a specific template. This can be used to introduce specific changes, including entire protein-coding domains such as split GFP into a gene, allowing endogenous tagging of proteins, or introduction of specific point changes.

This differs to approaches involving expressing genes on plasmids transfected into cells, because the host genomic changes are stable and heritable,

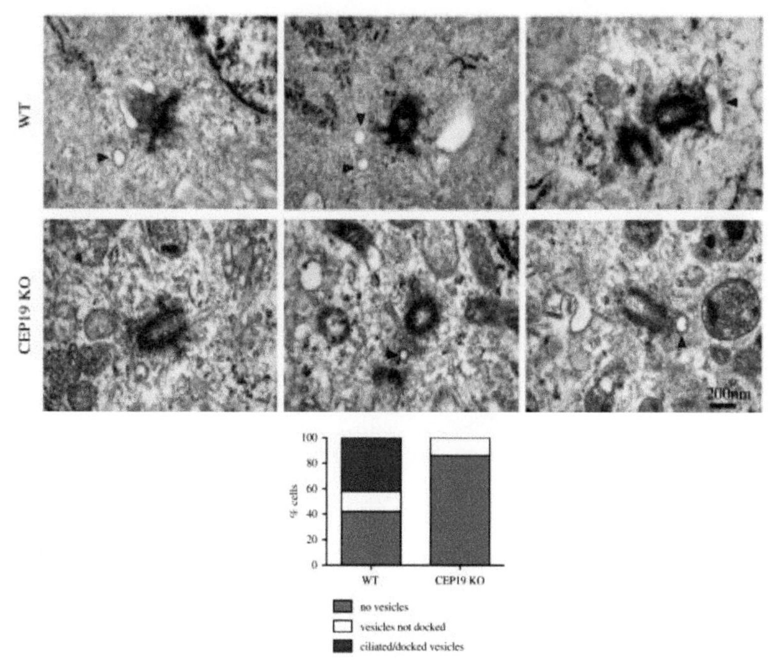

Figure 2.13 Transmission electron micrographs showing serial sections of wild-type (WT) and CEP19 knockout (CEP19 KO) RPE-1 cells, produced using CRISPR/Cas9 genome editing. Arrowheads indicate ciliary vesicles. Quantification of percentage of cells with docked/undocked ciliary vesicles and no vesicles in WT and CEP19 KO RPE-1 cells. Modified from reference [37]. Reproduced with kind permission of Prof. Laurence Pelletier.

and expressed at an endogenous level, in the fashion that the gene would normally be expressed.

Nucleases which require engineering to cut specific sequences include Zn finger nucleases [31] and TALENs [32]. The explosion in genome editing in the past 5 years has occurred as a result of the advent of the CRISPR/Cas9 approach [33, 34] which uses specific guide RNAs to target a common nuclease (Cas9) to the genomic region of interest. This requires significantly less molecular cloning, and has the advantage of being simple, rapid, and highly efficient.

A development of this technology, base editing, uses a cytidine deaminase enzyme tethered to inactive Cas9 to introduce specific single base changes, without introducing DSBs. This provides a fine detailed approach to genome editing [35, 36].

These approaches have been effectively used to genetically modify ciliated cells, from studies of individual genes (Figure 2.13) [37], through to large-scale screens of thousands of genes [38].

Acknowledgments

GW is funded by National Eye Research Centre, Wellcome Trust and UWE Bristol Quality Research funds.

References

[1] Kim J, Lee JE, Heynen-Genel S, et al. Functional genomic screen for modulators of ciliogenesis and cilium length. *Nature*, 464(7291), 1048-U114, 2010.

[2] Wheway G, Schmidts M, Mans DA, et al. An siRNA-based functional genomics screen for the identification of regulators of ciliogenesis and ciliopathy genes. *Nat. Cell Biol.* 17:1074–1087, 2015.

[3] Horn T, Arziman Z, Berger J, and Boutros M. GenomeRNAi: a database for cell-based RNAi phenotypes. *Nucleic Acids Res.* 35:D492–7, 2007.

[4] Arnaiz O, Malinowska A, Klotz C, et al. Cildb: A knowledgebase for centrosomes and cilia. *Database: The Journal of Biological Databases and Curation*, 2009:bap022, 2009.

[5] Arnaiz O, Cohen J, Tassin AM, et al. Remodeling cildb, a popular database for cilia and links for ciliopathies. *Cilia,* 3(1):9, 2014.

[6] van Dam TJ, Wheway G, Slaats GG, et al. The SYSCILIA gold standard (SCGSv1) of known ciliary components and its applications within a systems biology consortium. *Cilia*, 2(1):7-2530-2-7, 2013.

[7] Meier KA, and Insel P. "Hormone receptors and response in cultured renal epithelial cell lines," in M. Taub (Ed.), *Tissue Culture of Epithelial Cells*, 145–178, 1985.

[8] Ott C, and Lippincott-Schwartz J. Visualization of live primary cilia dynamics using fluorescence microscopy. *Current Protocols in Cell Biology, Chapter 4*, Unit 4.26, 2012.

[9] Gerdes JM, Liu Y, Zaghloul NA, et al. Disruption of the basal body compromises proteasomal function and perturbs intracellular wnt response. *Nature Genetics*, 39(11):1350–1360, 2007.

[10] Kowal TJ, and Falk MM. Primary cilia found on HeLa and other cancer cells. *Cell Biology International*, 39(11):1341–1347, 2015.

[11] Abbondanzo SJ, Gadi I, and Stewart CL. Derivation of embryonic stem-cell lines. *Methods in Enzymology*, 225:803–823, 1993.

[12] Villegas J, McPhaul M. Establishment and culture of human skin fibroblasts. *Current protocols in molecular biology*, John Wiley & Sons, 2001.

[13] Firth AL, Dargitz CT, Qualls SJ, et al. Generation of multiciliated cells in functional airway epithelia from human induced pluripotent stem cells. In *Proceedings of the National Academy of Sciences of the United States of America*, 111(17):E1723-30, 2014.

[14] Langousis G, and Hill KL. Motility and more: The flagellum of trypanosoma brucei. *Nature Reviews Microbiology*, 12(7):505–518, 2014.

[15] Harris EH, Witman GB, and Stern D. *The Chlamydomonas Sourcebook, Second Edition, Volumes 1–3 (2nd ed.)*. (Oxford: Academic Press, Elsevier), 2009.

[16] Vincensini L, Blisnick T, and Bastin P. 1001 model organisms to study cilia and flagella. *Biology of the Cell*, 103(3):109–130, 2011.

[17] Ormo M, Cubitt AB, Kallio K, et al. Crystal structure of the aequorea victoria green fluorescent protein. *Science*, 273(5280):1392–1395, 1996.

[18] Leonetti MD, Sekine S, Kamiyama D, et al. “A scalable strategy for high-throughput GFP tagging of endogenous human proteins,” *Proceedings of the National Academy of Sciences of the United States of America*, 113(25):E3501-8, 2016.

[19] Kamiyama D, Sekine S, Barsi-Rhyne B, et al. Versatile protein tagging in cells with split fluorescent protein. *Nature Communications*, 7:11046, 2016.

[20] Deisseroth K. Optogenetics: 10 years of microbial opsins in neuroscience. *Nature Neuroscience*, 18(9):1213–1225, 2015.

[21] Jansen V, Alvarez L, Balbach M, et al. Controlling fertilization and cAMP signaling in sperm by optogenetics. *eLife*, 4:10.7554/eLife.05161, 2015.

[22] Ryu MH, Moskvin OV, Siltberg-Liberles J, et al. Natural and engineered photoactivated nucleotidyl cyclases for optogenetic applications. *The Journal of biological chemistry*, 285(53):41501–41508, 2010.

[23] Stierl M, Stumpf P, Udwari D, et al. Light modulation of cellular cAMP by a small bacterial photoactivated adenylyl cyclase, bPAC, of the soil bacterium Beggiatoa. *The Journal of Biological Chemistry*, 286(2):1181–1188, 2011.

[24] Wyart C, Del Bene F, Warp E, et al. Optogenetic dissection of a behavioural module in the vertebrate spinal cord. *Nature*, 461(7262): 407–410, 2009.

[25] Roux KJ, Kim DI, Burke B. BioID: a screen for protein-protein interactions. *Current Protocols in Protein Science*, 74:Unit 19.23, 2013.

[26] Gupta GD, Coyaud E, Goncalves J, et al. A Dynamic Protein Interaction Landscape of the Human Centrosome-Cilium Interface. *Cell*, 163(6):1484–1499, 2015.

[27] Boldt K, van Reeuwijk J, Gloeckner CJ, et al. Tandem affinity purification of ciliopathy-associated protein complexes. *Methods in Cell Biology*, 91:143–160, 2009.

[28] Texier Y, Toedt G, Gorza M, et al. Elution profile analysis of SDS-induced subcomplexes by quantitative mass spectrometry. *Molecular and Cellular Proteomics*, 13(5):1382–1391, 2014.

[29] Boldt K, Gloeckner CJ, Texier Y, et al. Applying SILAC for the differential analysis of protein complexes. *Methods in Molecular Biology (Clifton, NJ)*, 1188:177–190, 2014.

[30] Boldt K, van Reeuwijk J, Lu Q, et al. and UK10K Rare Diseases Group. An organelle-specific protein landscape identifies novel diseases and molecular mechanisms. *Nature Communications*, 7:11491, 2016.

[31] Bibikova M, Beumer K, Trautman JK, et al. Enhancing gene targeting with designed zinc finger nucleases. *Science,* 300:764, 2003.

[32] Wood AJ, Lo TW, Zeitler B, et al. Targeted genome editing across species using ZFNs and TALENs. *Science*, 333(6040):307, 2011.

[33] Cong L, Ran FA, Cox D, et al. Multiplex genome engineering using CRISPR/Cas systems. *Science*, 339(6121):819–823, 2013.

[34] Mali P, Yang L, Esvelt KM, et al. RNA-guided human genome engineering via Cas9. *Science*, 339(6121):823–826, 2013.

[35] Gaudelli NM, Komor AC, Rees HA, et al. Programmable base editing of A*T to G*C in genomic DNA without DNA cleavage. *Nature*, 551(7681):464–471, 2017.

[36] Komor AC, Kim YB, Packer MS, et al. Programmable editing of a target base in genomic DNA without double-stranded DNA cleavage. *Nature*, 533(7603):420–424, 2016.

[37] Mojarad BA, Gupta GD, Hasegan M, et al. CEP19 cooperates with FOP and CEP350 to drive early steps in the ciliogenesis programme. *Open Biology*, 7(6):170114, 2017.

[38] Pusapati GV, Kong JH, Patel BB, et al. CRISPR Screens Uncover Genes that Regulate Target Cell Sensitivity to the Morphogen Sonic Hedgehog. *Developmental Cell*, 44(1):113–129.e8, 2018.

3

Programming Ciliary Object Manipulation

Richard Mayne

Unconventional Computing Laboratory, University of the West of England, Bristol BS16 1QY, United Kingdom
E-mail: Richard.Mayne@uwe.ac.uk

We are aware that laboratory experimental cilia research will lead to groundbreaking discoveries in the fields such as human medicine, engineering of "smart surfaces" and bio-inspired algorithm design, but we are still limited by our incomplete understanding of the basic control of biological cilia activity by live cells. In this chapter, we examine studies that aimed to investigate the natural ability of a ciliated model organism, *Paramecium caudatum*, for completing a task with demonstrable practical use: manipulation of environmentally dispersed particulates by the organisms' cilia. We then examine consequent studies that were designed to "hijack" the organisms' ciliary dynamics in a non-invasive manner toward harnessing this desirable behavior. It is found that *Paramecium* is able to differentially manipulate ("sort") particles with their cilia on the basis of size, by both active and passive mechanisms. While it is experimentally challenging to control individual cilia in a live organism, influencing global cell behavior through facile methods is a viable route toward achieving demonstrably "useful" goals if the experimental environment capitalizes on said organisms' natural behaviors, in this case, migratory and feeding mechanisms powered by cilia. We conclude by discussing how cilia dynamics can be understood in the language of biocomputation and how in useful devices powered by cilia may be built and "programmed".

3.1 Introduction

It is a fundamental concept in the field of unconventional computing that all matter is programmable, at least hypothetically. Any physical quantity may be interpreted as data and interactions between these quantities that are put to some practical use constitute computation. Programming in this sense is therefore the assignation of logical, coherent input and output parameters to data, prior to engineering the conditions by which computation will occur (operation). Unconventional computing's daughter field, biocomputing, is no exception to this rule, although the inherent complexity, massive parallelism, and partially unelucidated nature of biological systems prevent us from exerting full control over their machinery. This does not mean that laboratory experimental biocomputer prototypes are impossible to make and manipulate toward some demonstrable purpose however, as live organisms possess the ability to self-assemble their own functional units that have their own in-built parameters of operation. The motile cilium is one such functional unit.

The benefits of reprogramming multiciliated cells – be it with genetic, electrical, optical, tactile, or chemical stimuli – are threefold. First, a significant subset of human diseases are associated with either acquired or inherited ciliary defects (see Chapter 2 for an overview of this topic), implying that overhauling these cells' ciliary machinery in a repeatable, programmable manner is a viable route toward novel therapies. Second, as experimental manipulation of ciliary dynamics necessitates that a certain knowledge of the system is first gained, research in the field will inform the creation of artificial, "biomimetic" cilia which capitalize upon (and possibly even enhance) the natural functions of cilia using novel materials; the applications for such technologies are virtually limitless and significant steps toward this have already been made, see Refs. [1–7]. Third, exerting full control over the cilia arrays possessed by non-mammalian cells, such as the ciliates (single-celled multiciliated protozoa), presents a unique opportunity for manipulating aquatic environments at the micro-, or even nano-scale, e.g., as a method for "cleaning" bodies of water contaminated with environmental pollutant particulates [8].

It was toward all of these goals that the experiments detailed here were addressed during the course of a Leverhulme Trust-funded grant, which aimed to implement programmable sorting of micro-scale objects by live ciliated cells – the model organism *Paramecium caudatum* (Figure 3.1) – as well as fabricate a meso-scale *P. caudatum* cilia-inspired manipulator robot

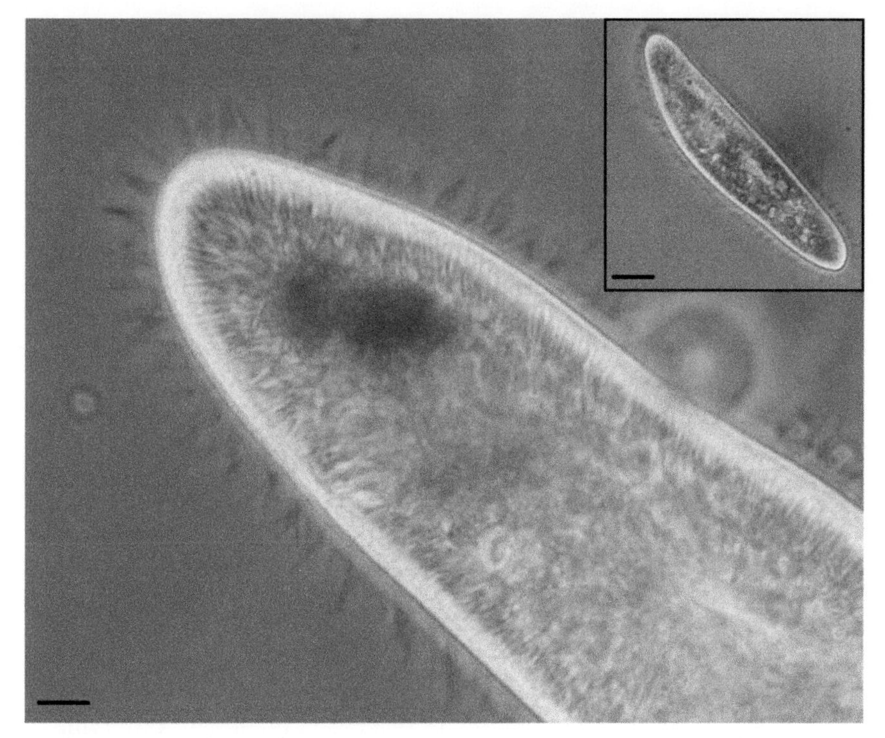

Figure 3.1 Phase contrast photomicrographs showing *P. caudatum* cell. (Main) Anterior tip of the cell, showing hundreds of cilia protruding from the organism's surface. Scale bar 10 μm. (Inset) Lower magnification image showing the whole cell. Scale bar, main, 10 μm, inset, 25 μm.

also capable of sorting. This chapter details findings of the former aspect of this study and Chapter 4 examines the latter.

3.2 Sorting by Paramecium Cilia

"Sorting" in this context refers to differential manipulation of objects based on one or more physical property, be it size, chemical constituents or color. The process, although distinct from the mathematical description of sorting, implies an algorithmic two-stage process comprising sensing (determination of object properties) and actuation (physical manipulation of said object).

Particulate manipulation by *P. caudatum* is a well-observed phenomenon: as obligate filter feeders, they migrate toward attractant chemical gradients wherein dispersed foodstuffs – bacteria, yeasts, and smaller protozoa – are directed by specialized cilia toward a mouth-like cavity called the cytopharynx. The various cytopharyngeal cilia arrays beat independently of the organism's somatic cilia at a higher frequency and serve to concentrate foodstuffs within a specialized phagocytotic vesicle at the cavity terminus through which they are ingested, while ejecting excess fluid and particulates that are not ingested [9]. Contemporary thought states that the manner in which oral cilia interact with edible particulates is entirely passive, acting simply as a "sieve" [9–11]. This process constitutes a form of sorting in its own right: although the organism does the job passively, it nevertheless constitutes active discrimination between objects. This is an interesting distinction when contemplated in relation to a recent movement in the field of robotics, "morphological computation," which emphasizes that computing done passively as a product of an entity's morphology (physical dimensions and assembly) should be capitalized upon to the fullest extent toward reducing complexity while increasing efficiency [12]. This is, of course, a bio-inspired development.

The ultrastructure of a *P. caudatum* oral cilium is essentially identical to a somatic cilium; their principal difference is simply location, hence their local environmental conditions and possibly internal linkage differ. This raises the important question as to how can the organism, as a functional unit, manipulate particulates using its full complement of cilia, rather than just those in its mouth. Our first series of investigations [13] focused on simply observing how *P. caudatum* interacts with dispersed particulates in its environment based on their size, as a necessary precursor before investigating methods of their experimental manipulation. Experiments consisted of cultivating the organisms in the presence of fluorescent latex particles of a range of sizes (0.2, 2.0, and 15 μm) and observing their movements in real time using fluorescence microscopy (Figure 3.2). Movements (trajectory, velocity) of particulates were analyzed with custom software.

We identified two dominant modes of organism–particulate interaction that appeared to be dependent on the momentary state of the organism, which were classified as being either "sessile" or "motile" (Figure 3.3). The sessile behavior pattern, which was characterized by little or no directional migration, exhibited a characteristic set of fluid vortices that drew the two smaller varieties of particulate toward the oral apparatus and cycled them over the region repeatedly – presumably as a means of enhancing feeding. Conversely, the motile state was characterized by direct and rapid directional migration of the organism, during which the pattern of particle movements was characterized by the smaller two sizes of parting being drawn across the organism and rapidly ejected in a fluid contrail. The larger variety of particles

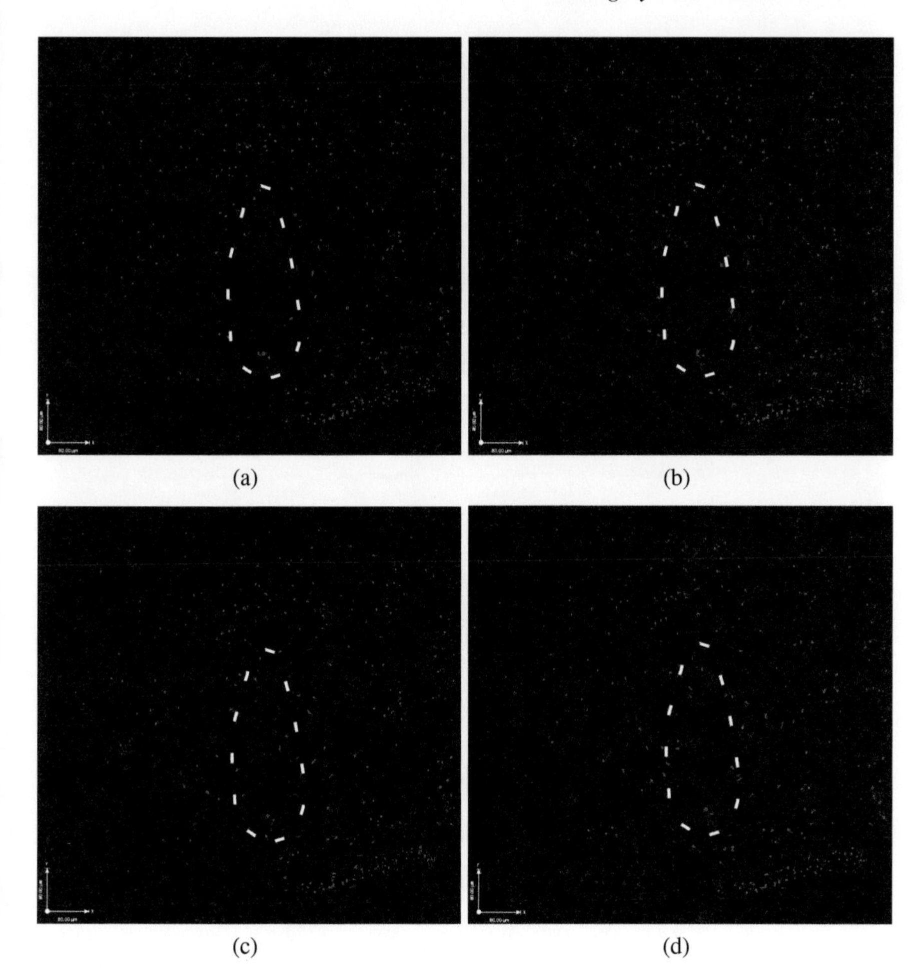

Figure 3.2 Sequential fluorescence micrographs (a–d, 35 ms intervals) of a *P. caudatum* cell creating fluid flows about its body consistent with "sessile behavior", via beating of its cilia. Movements of the adjacent fluorescent 2.0 μm latex particles demonstrate the fluid vortices created. The cell's outline is highlighted. Reproduced from [13].

were completely unaffected by cilia-mediated fluid currents in the organisms' motile phase, but were slowly drawn into close proximity with the organisms during the sessile phase. In these instances, the larger particulates were held static at a range of about one cilium's length (10 μm) from the organism for indefinite periods of time. We hypothesized that this was due to the creation of an electrostatic double layer which permits the immobilization of larger objects while smaller particulates may be stripped off them (e.g., decaying

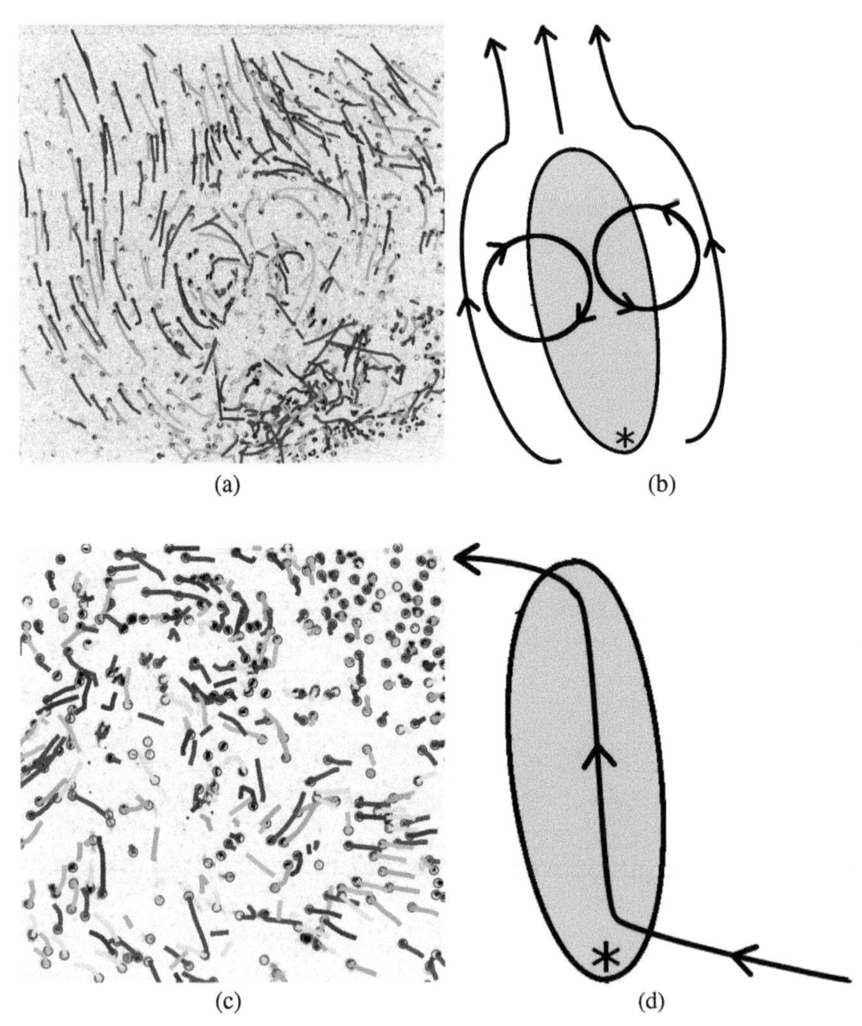

Figure 3.3 Fluid currents about (a and b) sessile [output from Figure 3.2] and (c and d) motile *P. caudatum* cells, as demonstrated by the movements of cilia-driven fluorescent 2.0 μm latex spheres. (a and c) Output from particle tracking script where particles are shown as black dots with multi-colored tails. (b and d) Diagrams of cells showing directions of fluid current. Asterisks indicate the organisms' anterior tip. Reproduced from [13].

organic detritus in their natural habitat), but the phenomenon requires further research before any conclusions may be drawn.

Our conclusions were, therefore, that *P. caudatum* possesses the means for actively discriminating between particulates of different sizes and is therefore capable of "sorting", apparently via both active (changing swimming dynamics to enhance feeding on smaller particles) and passive (immobilization of bigger particles while drawing in smaller ones) modes. We did not observe any significant changes in ciliary beating frequency between different modes, which persisted at around 30 Hz with only minor deviations; metachronal wavelength and velocity were both highly variable between experiments and no correlations were made, although the microscopical techniques used precluded any forms of three-dimensional measurement that could have appropriately tracked metachronal wave propagation along the helical paths they follow in *Paramecium* species.

Crucially, *P. caudatum* was found to ingest virtually everything ranging from 0.2 to 2.0 μm, including fluorescent latex spheres and a range of metallic nanoparticles, the latter of which were used in a companion study investigating the toxic effects of various nanomaterials on the organism (Figure 3.4) [8].

3.3 Reprogramming Paramecium Cilia

Initial results toward making a "useful" cilia biocomputer revealed that manipulating *P. caudatum* cilia individually was beyond the limitations of current technology, despite some successes in manually changing ciliary beating frequency and metachronal wave characteristics through the use of multisensorial stimulation (optical, chemical, and electrical) on organisms immobilized in glass needles (Figure 3.5). More specifically, we were able to demonstrate that favorable chemical gradients would upregulate beating, strong optical gradients would cause the inverse, and DC electrical fields could be fine-tuned to produce either effect [14], the mechanisms for which correspond to the organism's taxes in response to these stimuli.

Treating a live biological substrate as a linear series of input and output relations is somewhat of a waste of their true potential. Thoughtful biocomputer prototypes capitalize on existing hardware and disrupt the cellar environment as little as possible. We proceeded to build a proof-of-concept device demonstrating orchestrated manipulation of nano-scale objects by *P. caudatum* cells; the functional units of these devices were cilia, but they were "operated" by the cells and only 'programmed' by us. Our experiments were designed to create a parallel, cilia-driven nano-object manipulator that could be tasked to clear a three-dimensional region of fluid of contaminant particles. A basic variant of the devices created is shown in Figure 3.6: they were, very simply, a series of chambers connected by small PTFE tubes.

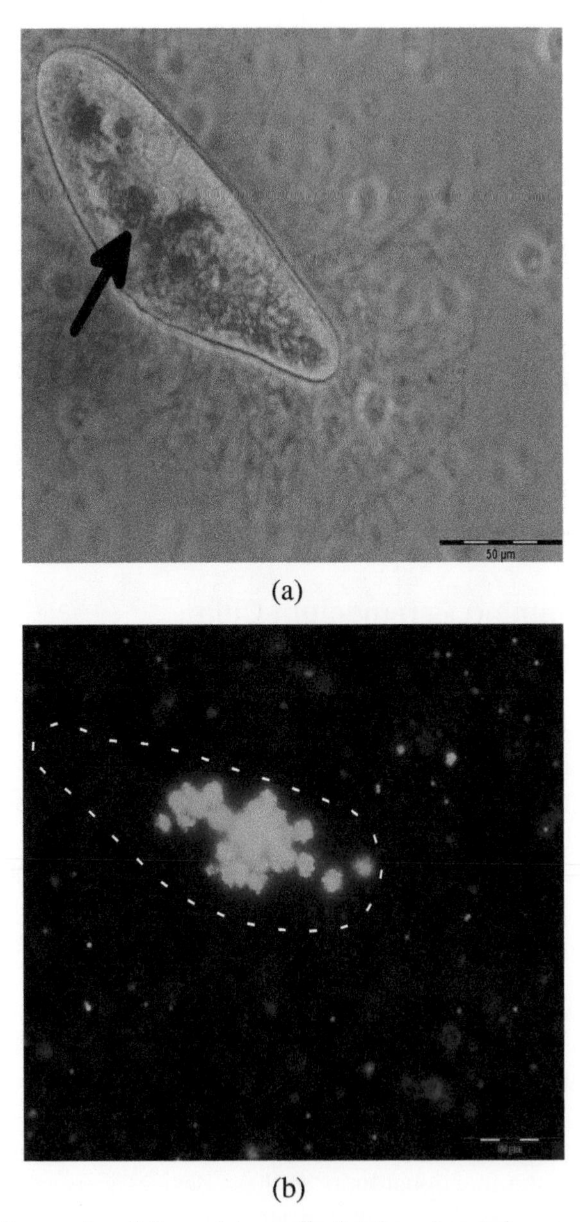

Figure 3.4 Micrographs of *P. caudatum* cells that have ingested exogenous nano- and micro-scale materials. (a) Light micrograph of a cell that has ingested a quantity of 200 nm magnetite nanoparticles, which present as rust-colored cytoplasmic inclusions (arrowed). (b) Fluorescence micrograph of a cell following ingestion of 2.0 μm fluorescent latex spheres, which are also contained in spherical masses within the cell. The cell's outline is highlighted. Copyright Jack Morgan 2018.

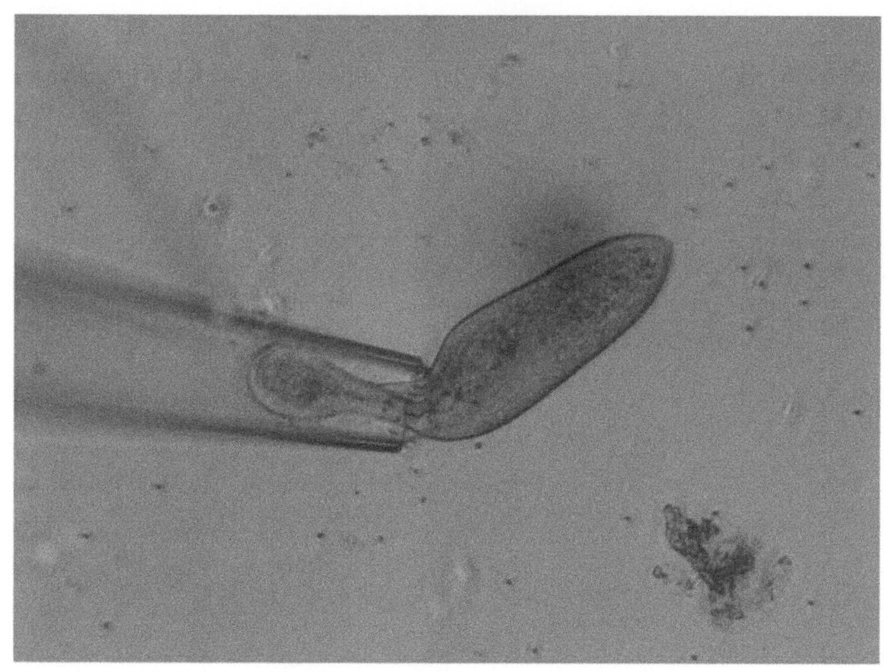

Figure 3.5 Micrograph to show a *P. caudatum* cell immobilized within the tip of a hollow glass needle. This experimental setup was used to investigate cilia dynamics in response to a variety of stimuli. For scale, the tip of the needle is 75 μm. Original magnification $\times$100. Copyright Richard Mayne 2018.

P. caudatum cells were placed in one chamber and would migrate toward other chambers that were contaminated with nanoparticles as a result of being guided there by supplementary chemoattractants (starch). Once they had migrated toward the supplied attractant, the cells would proceed to ingest quantities of the contaminants – in this case, magnetite nanoparticles – due to their lacking the ability to discriminate between particles of this particular size. By consequently engaging repellent optical or electrical gradients in the contaminated chambers, the cells could then be diverted back to their original chamber, wherein they could be removed or decellularized in order to free the internalized particles. This guided cell movement was found to be repeatable and reliable, and did not damage the cells.

We demonstrated that these biohybrid devices are a viable route toward programmable ciliary manipulation, which we consequently automated via the introduction of some basic low-cost machinery for fluorescence spectrometry (Figure 3.7). Our consequent experiments continue at the time

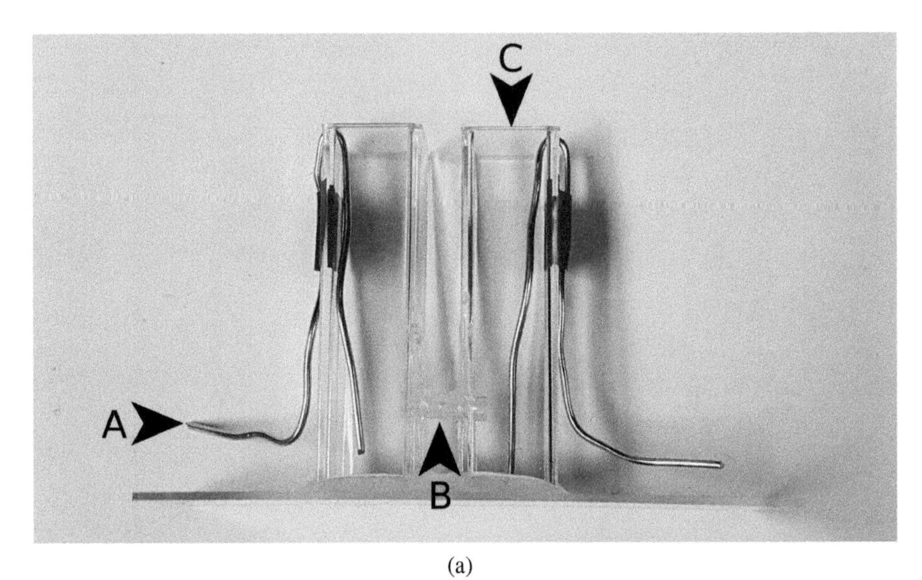

(a)

(b)

Figure 3.6 Photographs of programmable *P. caudatum*-operated nano-object manipulator. (a) Experimental environment. A: platinum electrode, B: 1 mm diameter tube connecting chambers, C: chamber, fashioned from a standard fluorescence cuvette. The environment is affixed to the base of a 90 mm microscope slide. (b) Environment mid-experiment (electrodes removed). Cells are placed in chamber A and a mixture of nanomaterials and chemical attractants are placed in chamber B, encouraging their migration $A \rightarrow B$. Then, electrical or optical input may be used to guide the cells' migration back $B \rightarrow A$. Reproduced from [14].

Figure 3.7 Photograph of DIY fluorescence spectrometer attached to an experimental environment chamber, for automating the output of a *P. caudatum*-driven nano-manipulator. Reproduced from [14].

of writing into evaluating the efficiency and range of uses of such devices in environmental management settings. Although this is possibly far from the classical concept of computation, we nevertheless find that the output of these "machines" may be equated with computer logic. For instance, the presence of contaminants in a specific chamber may be interpreted as a logical TRUE. This enhances automation of output recognition. Although we used spectrometry as a biological-computer interface, the concept presents fascinating potential technologies for cilia-based manipulators, especially in the field of enhancing the actions of motile cilia in human medicine, provided that optical interfaces for automatic interpretation of ciliary beating are developed.

References

[1] Sareh S, Rossiter J, Conn A, et al. Swimming like algae: biomimetic soft artificial cilia. *Journal of the Royal Society Interface*, 10(20120666), 2012.

[2] Vilfan M, Potocnik A, Kavcic B, et al. Self-assembled artificial cilia. *Proceedings of the National Academy of Sciences of the United States of America*, 107(5):1844–1847, 2010.

[3] Georgilas I, Adamatzky A, Barr D, et al. Metachronal waves in cellular automata: Cilia-like manipulation in actuator arrays. In *VI International Workshop on Nature Inspired Cooperative Strategies for Optimization (NICSO 2013)*, 2013.

[4] Zhou Z, and Liu Z. Biomimetic cilia based on MEMS technology. *Journal of Bionic Engineering*, 5(4):358–365, 2008.

[5] den Toonder JM, and Onck PR. Microfluidic manipulation with artificial/bioinspired cilia. *Trends in Biotechnology*, 31(2):85–91, 2013.

[6] Assaf T, Mayne R, Adamatzky A, et al. Emergent behaviors in a bioinspired platform controlled by a physical cellular automata cluster. *Biomimetics*, 1(5):1–13, 2016.

[7] van Oosten CL, Bastiaansen CWM, and Broer DJ. Printed artificial cilia from liquid-crystal network actuators modularly driven by light. *Nature materials*, 8(8):677–682, 2009.

[8] Mayne R, Whiting J, and Andrew A. Toxicity and applications of internalised magnetite nanoparticles within live *Paramecium caudatum* cells. *BioNanoScience*, 2017.

[9] Ishida M, Allen R, and Fok AK. Phagosome formation in Paramecium: roles of somatic and oral cilia and of solid particles as revealed by video microscopy. *Journal of Eukaryotic Microbiology*, 48(6):640–646, 2001.

[10] Ramoino P, Diaspro A, Fato M, et al. Imaging of endocytosis in Paramecium by confocal microcopy. In: *Molecular Regulation of Endocytosis*, pages 123–152, 2012.

[11] Verni F, and Gualtieri P. Feeding behaviour in ciliated protists. *Micron*, 28(6):487–504, 1997.

[12] Hauser H, Ijspeert AJ, Fuchslin RM, et al. Towards a theoretical foundation for morphological computation with compliant bodies. *Biological cybernetics*, 105:355–370, 2011.

[13] Mayne R, Whiting JGH, Wheway G, et al. Particle sorting by Paramecium cilia arrays. *Biosystems*, 156–157:46–52, 2017.

[14] Mayne R, Morgan J, Phillips N, et al. Programmable transport of micro- and nanoparticles by *Paramecium caudatum*. *BiorXiv Preprint*, 2018.

PART II

Engineering

4

Robotic Cilia for Autonomous Parallel Distributed Sorting Platforms

James G. H. Whiting

Unconventional computing Group, Frenchay Campus,
University of the West of England, Bristol BS16 1QY, United Kingdom
E-mail: James.Whiting@uwe.ac.uk

Abstract

Industrial object manipulation is traditionally performed by conveyor systems and robotic armatures. Each of these technologies is traditionally controlled centrally via a CPU or control system. We have developed and demonstrated an object manipulation system capable of differentially sorting objects which uses a system of parallel distributed control. Each component in the system has its own control mechanism or sensor–microcontroller–actuator; importantly, the system is able to organize object movement by communicating with its neighbors to map its local environment and perform object sorting and manipulation tasks. The sensing and actuation system is inspired by cilia on *Paramecia caudatum*. Robotic armatures mimic the ciliary action and sensing elements on each cilium detect the type of object above them. We have demonstrated sorting objects based on color and rotation using edge detection; this proof-of-concept system is largely up-scalable and reprogrammable.

4.1 Introduction

Object manipulation is possible using several methods: conveyor belts, vibratory bowl feeders, and combinational technology. A lot of sorting systems use open loop or very low level control systems to sort objects; mail sorting centers use a combination of conveyor belt systems and address reading systems to sort a very high volume of letters and parcels. Similarly, recycling centers use comparable technology with optical sensor systems to sort recycling into several recyclable categories. There has been an interest in replicating cilia in hardware in order to produce a unidirectional platform which could replace conveyor belts and other object manipulation technology. Micro electrical mechanical systems (MEMS) have been developed which manipulate small objects (several millimeters in size) such as microchips or other electronic components [1]. Systems such as MEMS comprise four vibrating or oscillating arms which manipulate objects by vibrating combinations of these arms [2]. While they are claimed to be biomimetic cilia, they do not mimic the ciliary beating action, nor do they physically resemble cilia [3]. Cilia themselves are used to create water currents, for either propulsion or feeding of aquatic cellular organisms, or in multi-celled organisms as manipulation of adjacent fluids and objects. Rarely do they manipulate objects outside of an aqueous environment. That said, one group has produced an artificial cilium using soft robotic actuators, which very closely mimics the ciliary beat of a cilium. This cilium can be used to create water currents; however, our task is somewhat different. Another aspect of artificial cilia platforms is that they are usually controlled by one central processor, with feedback given by way of an overhead camera. Platforms such as this have limitations when it comes to changing the layout of the platform or scaling the size. Simultaneous cilia are used in a platform in order to manipulate objects over a surface; these ciliary units are controlled in parallel which again causes issues with scalability.

Distributed intelligence in sorting platforms and indeed other multinodal systems allows for the removal of a central control unit; emergent properties arising from a simple set of behavioral rules combined with distributed control can create apparent intelligence in very simple systems such as swarms or single-celled organisms. We have developed a system, which has no central processor for control; instead, it relies on distributed control, each cilium unit is capable of operating independently.

In order to allow for significant cooperation and sorting, we have implemented a hexagonal cell structure, similar to that of real cilia in a cell membrane (Figure 4.1); each cilium is capable of communicating with its six neighbors. The organism's membrane holds the cilia in alignment in the continuous outer membrane; under sufficient magnification, a hexagonal tessellation of longitudinal rows is visible, with one cilium protruding from each hexagonal structure. It is because of this hexagonal tessellation that each ciliary unit's circuit board has hexagonal outer dimensions. This tessellation provides opportunities for sharing power, providing a system which is entirely

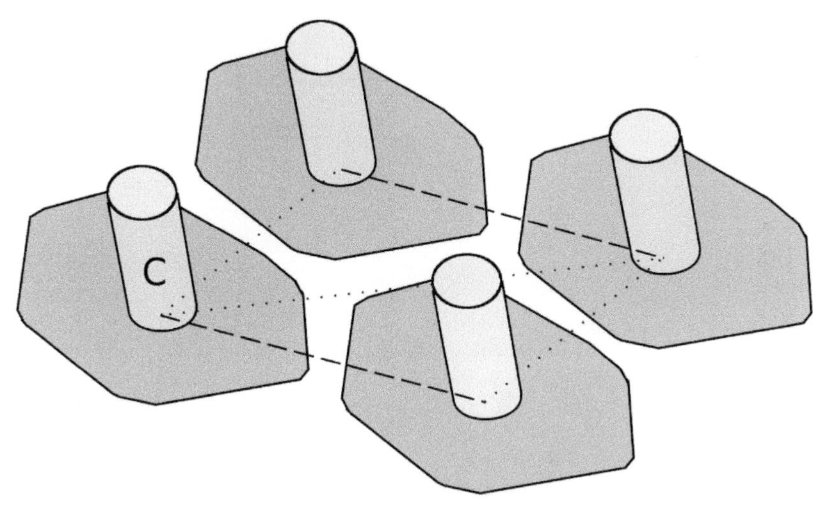

Figure 4.1 A diagrammatic representation of the cilia, c, protruding from the cell membrane, with interconnections of kinetodesmal fibrils and infraciliature, in hexagonal structure. Copyright JGH Whiting.

Figure 4.2 Vibrating motor platform showing distributed control of eight sensor–actuator pairings controlled by one microprocessor. Each microprocessor uses local communication between its neighboring microprocessors to organize bidirectional movement of objects placed on the platform. Bi-directionality is obtained by rotating the motors either forward or backward. Copyright JGH Whiting.

Figure 4.3 Biomimetic cilium in hexagonal arrangement, which tile together like biological cilia. Three angles of view are visible in this image (sagittal, coronal, and transverse planes) in order to demonstrate the three-servo robotic armature used to mimic the ciliary beat. A color sensor is placed on top of the distal servo and used to identify objects placed on the platform; in this application, color is used to sort objects in three different directions. Local power sharing and bidirectional data are sent using the five header pins on each edge of the printed circuit board, connected by jumpers. Communication can be synchronous or asynchronous depending on if a master board is chosen. Copyright JGH Whiting.

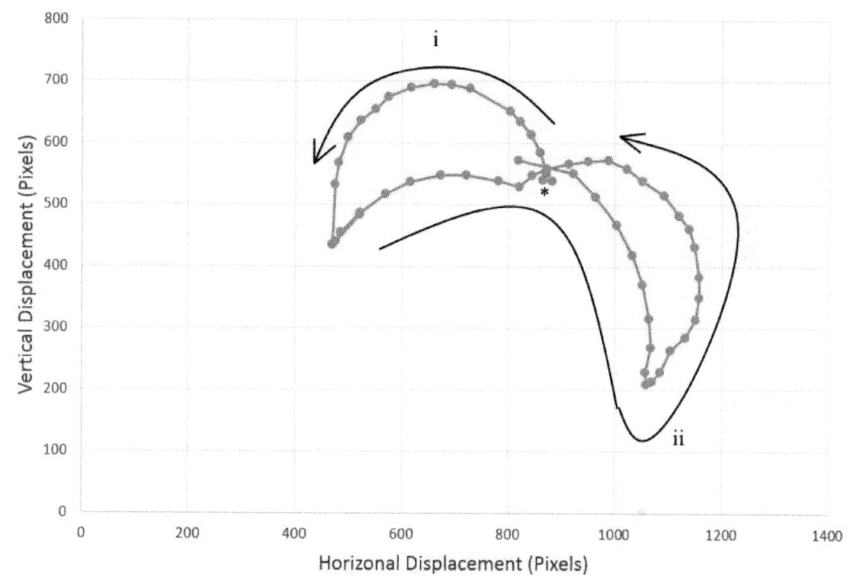

Figure 4.4 A two-dimensional plot of biomimetic robotic ciliary beating pattern using the servo system in Figure 4.3. The beat, starting at the marked *, performs the power stroke (i) which propels the object forward, followed by a recovery stroke, (ii) which returns the cilium tip to the original position without moving the object. While real cilia beat in three dimensions, having a more circular lateral movement, the side profile of cilium beat is fundamentally similar in this robotic armature to its biological counterpart. Copyright JGH Whiting.

re-configurable in morphology; additionally, it is significantly biomimetic in geometry. This local communication allows local clusters of cilia to coordinate action and act together. This degree of only local coordination produces complex behavior similar to that of swarms of birds or fish; fish swimming in shoals cannot possibly know where all the other fish in the shoal are; instead they concentrate on avoiding collision with neighboring fish while maintaining distance. Local communication like this creates swarms of fish which group together for protection and move as one giant structure without one animal in charge.

Figure 4.5 Cilia hexagonal tiling demonstrated using 27 tiled boards sharing power and communicating with neighbors. Objects were manipulated in different directions along the entire platform. Copyright JGH Whiting.

4.2 Engineering Sorting Ciliary Platforms

We initially developed a prototype platform based on a cellular automata (CA) platform; CAs also use local communication and create complex patterns (some of which can be seen in nature, such as in the shell of some species of cone snail). Like the emergent intelligence of the paramecia, a simple set of rules governs the organism's response, leading to apparently complex behavior. Adapting this platform, we added vibrating motors (Figure 4.2) which, much like previous vibratory bowl feeders and vibrating actuators, can manipulate objects. These motors are bi-directional and therefore offer two directions of manipulation. These platforms share power and are hugely tileable (subject to power constraints); so in an attempt to increase directionality of the surface and increase the biomimetic proportion of the cilia actuator, we redesigned our own cilia actuators and tileable boards. These manipulated objects and offered distributed control; however, their cilia-like action was limited.

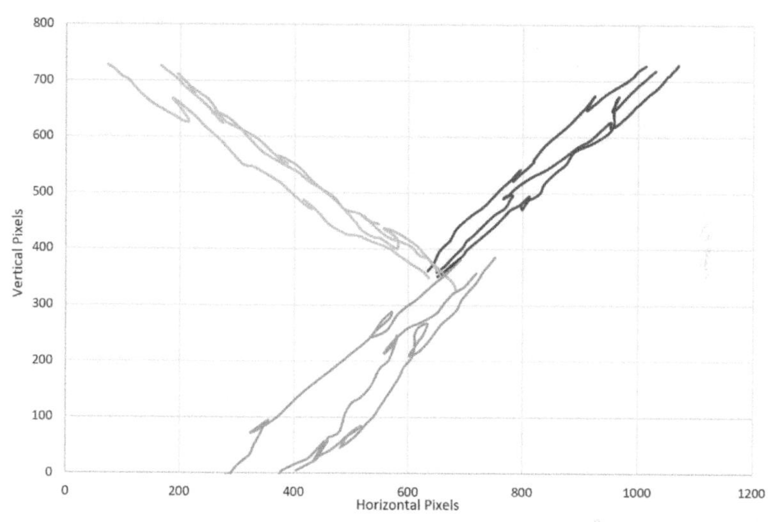

Figure 4.6 Directional movement of different colored objects represented by the corresponding color line. Using red, green, and blue objects, the platform was able to move each object correctly and at a speed of up to 4 cm/s. Copyright JGH Whiting.

To create biomimetic cilia movement, we produced a hinged servo robotic arm, consisting of three servos [4]. Three Tower Pro SG50 Micro Digital Servos (Servo Shop, UK) were used to copy the whipping power and recovery stroke, with two servos mounted axially to create a two-segment robotic arm, and the third servo placed orthogonally at the base of the arm, to facilitate rotation of the cilium to stroke in different directions (Figure 4.3). As the focus of this work is to create a bioinspired sorting platform, a variety of sensors could be used, dependent on the practical application, of which there is a huge variety; the microcontroller can interact with a huge array of analog or digital sensors, so desired sensing systems can be easily integrated for each application. The three-servo system allows the robotic cilia to beat and mimic the power and recovery stroke of biological cilia (Figure 4.4). A color sensor is placed on top of each cilium, so that it may sense its own environment and share that information with its neighbors. The platform (Figure 4.5) was capable of coordinating the sorting of objects in different directions based on the object's color (Figure 4.6); the objects used to test this were small jiffy bags of 20 cm × 15 cm weighing 127 g (Figure 4.7). The velocity of the object was proportional to beat speed, so the speed of manipulation could be controlled; the fastest speed is 45.6 mm/s; there was no discernible difference between the speeds traveled by different colored objects.

Figure 4.7 Examples of sorting taken from the video feed used to analyses the speed of movement. The jiffy bags contain small electronic components from a supplier and weigh 127 g. The top image is taken from the beginning of the video when the item was placed on the platform, and the bottom image is taken from the end of the video when the object had reached the edge of the platform. Copyright JGH Whiting.

In order to detect and sort based on the object shape, the sorting platform must be able to identify different shapes. This can be most accurately performed with edge detection. As each cilium cell is able to communicate and share information with its local neighbor, it is able to determine which of its surrounding cells are "occupied". Using the ability to detect colored objects above its own headspace, each cell knows if an object is above it, and then by sharing this information with its neighbor, it is able to determine if it is part of an edge of an object or part of an inner cell. Cells with at least one or more neighbors whom are not "occupied," form part of the edge. If all cells in the platform perform this check, then the surface is able to identify the outline of an object. Rotation of an object is another task which can be performed by this platform. In order to perform rotation of an object about its origin, computation must be performed in order to determine either the outer edge or the center mass.

While we have maintained the objective of bio-mimicry, as with any bio-inspired work, which aims to derive a useful feature from nature's design, unless the hardware equivalent is near-identical to the natural counterpart in size, power, and task performed, then there has to be a tradeoff between similarity and engineering efficiency.

References

[1] Suh JW, Darling RB, Bohringer KF, et al. "Fully programmable MEMS ciliary actuator arrays for micromanipulation tasks," in *Proceedings 2000 ICRA. Millennium Conference. IEEE International Conference on Robotics and Automation. Symposia Proceedings (Cat. No.00CH37065)*, San Francisco, CA, vol. 2, 1101–1108, 2000.

[2] Zhou ZG, and Liu ZW. Biomimetic cilia based on MEMS technology. *Journal of Bionic Engineering*, 5(4):358–365, 2008.

[3] Whiting JGH, Mayne R, Meluish C, et al. A cilia-inspired closed-loop sensor-actuator array. *Bionic Engineering,* 1;15(3):526–532, 2018.

[4] Whiting JGH, Mayne R, Adamatzky A. A parallel modular bio-mimetic cilia sorting platform. *Biomimetics*, 3(2):5, 2018.

5

Artificial Pneumatic Cilia

Benjamin Gorissen[1,*], Edoardo Milana[1], Michaël De Volder[1,2] and Dominiek Reynaerts[1]

[1]Department of Mechanical Engineering, KULeuven, Celestijnenlaan 300, 3001 Heverlee, Belgium
[2]Institute for Manufacturing, University of Cambridge, 17 Charles Babbage Road, Cambridge CB3 0FS, UK
*Corresponding Author
E-mail: benjamin.gorissen@kuleuven.be; edoardo.milana@kuleuven.be; mfld2@cam.ac.uk; dominiek.reynaerts@kuleuven.be

5.1 Introduction

Well hidden from the macroscopic view of the human eye lays the fascinating word of low Reynolds fluid flow. This world is vastly different from the moderate to high Reynolds numbers that we encounter in our daily lives. Where the reciprocal motion of a fish's tail fin is an excellent mechanism for achieving fluid flow at $Re > 1$, a low Reynolds fish would not be able to advance using the same motion. Nature's evolutionary mechanism devised some unique solutions to the low Reynolds number fluid propulsion problem, such as vibrating hair-like structures that beat in an orchestrated pattern, called cilia. Due to the small dimensions of these cilia, they almost exclusively operate at low Reynolds number, where asymmetry is essential to achieve net propulsion. This is best known as the Scallop theorem that states that a low Reynolds swimmer must deform in a way that is not invariant under time reversal [1].

These asymmetrically moving cilia are difficult to mimic at a microscale using devices made by traditional fabrication processes. However, with the advent of micro- and nanotechnology, we now have a set of new tools to better imitate the propulsion of biologic microorganisms. The main challenge consists in developing artificial cilia that exhibit all types of asymmetries that natural cilia display, while having roughly the same size. In nature, four types of ciliary asymmetries are found. Three of them act on the level of a single cilium [2]: spatial asymmetry, where effective and recovery strokes do not follow the same path; orientational asymmetry, where the mean axis of the cilium is spatially tilted away from the surface normal; and temporal asymmetry where effective and recovery strokes have different velocities. At the level of the cilia array as a whole, additional asymmetry can be created by generating metachronal waves where a phase difference in actuation is present between neighboring cilia [3]. Although spatial asymmetry is a necessary boundary condition to achieve low Reynolds fluid flow, cilia in nature usually combine different types of asymmetries to form complex beating patterns in order to maximize fluid propulsion [4]. In the past decade, advanced artificial cilia were developed that exhibit multiple asymmetry modes, using different actuation technologies, with performances that match those of natural cilia. An overview of artificial cilia described in the literature can be found in Table 5.1, where the actuation principle is used to classify them, together with achieved asymmetry and cilium length. Since temporal asymmetry has no influence on low Reynolds fluid flow, it has been omitted in Table 5.1.

As can be concluded from Table 5.1, pneumatic artificial cilia are uniquely suitable to study the influence of different types of asymmetries on fluid flow, since all asymmetry types are reproducible and also adjustable, which is especially important in experimental investigations. This chapter will focus on pneumatically controlled artificial cilia, where two types of pneumatic cilia can be distinguished: those where passive hair-like structures are implanted on top of a pneumatically actuated base membrane and those

Table 5.1 Overview of artificial cilia in the literature

Actuation Principle	Asymmetry Type			Cilium Length (μm)	Reference
	Spatial	Orientational	Metachronal		
Magnetic	Yes	Yes	No	10–25	[5, 6]
	Yes	Yes	No	300	[7]
	Yes	Yes	No	31	[8]
	Yes	Yes	No	70	[3, 9, 10]
	Yes	Yes	No	300	[11]
	Yes	Yes	No	400	[12–14]
Electrostatic	No	Yes	Yes	100	[15]
Electroactive polymer	Yes	Yes	No	64	[16]
Optical	Yes	Yes	No	10,000	[17]
Base vibration	No	No	No	800	[18, 19]
Mechanical	Yes	No	Yes	500–1000	[20]
Pneumatic	Yes	Yes	Yes	500	[21, 22]
	No	Yes	Yes	10,000	[23]

where the body of the hair-like structure itself is pneumatically actuated. Both types will be described separately in the following paragraphs.

5.2 Base Actuated Pneumatic Cilia

A first type of artificial pneumatic cilia consists of passive platelets that are implanted at an angle on top of a flexible membrane, as shown in Figure 5.1, and as reported by Rockenback et al. [21, 22, 24]. Being plate-like in form and working at a Reynolds number equal to 5, these cilia mimic the propulsion system of ctenophores [25]. Each flap stands off-center on a thin membrane which can be pressurized or depressurized. This variation of pressure deflects the membranes changing their curvature, and consequently changing the inclination of the flaps on top. By imposing a sinusoidal pressure between −0.6 and 0.6 bar, the implanted plates exhibit spatial asymmetry, as can be seen in Figure 5.1d. Further, by changing the rate of inflation and deflation of the membrane, temporal asymmetry can be adjusted, and as consecutive membranes can be individually controlled, metachronal asymmetry is possible. As such, this system exhibits all four types of asymmetries, where metachronal and temporal asymmetry are adjustable by tuning the pneumatic input cycle, while orientational and spatial asymmetry are determined by the production process. The flaps, supporting membranes, and pneumatic connections are made out of PDMS using a soft lithography process involving two SU8 micromolds and a post-cure deformation step to introduce the inclination. Each flap measures 500 μm high and 50 μm thick, and is located on a membrane which is 175 μm thick and 600 μm wide. The pitch between the flaps is 1000 μm. The overall design is depicted in Figure 5.1b.

The performance of these base actuated plate-like cilia has been tested using experiments in water, where the effect of orientational asymmetry and metachronal asymmetry has also been studied. The influence of orientational asymmetry is quantified by using four different inclination angles of the flaps, while keeping metachronal, temporal, and orientational asymmetry constant. In all cases, a constant temporal asymmetry is achieved by imposing a positive actuation pressure for 40 ms (effective stroke) and a negative actuation pressure for 103 ms (recovery stroke). The symplectic metachronal asymmetry is set by having a fixed time shift of 20 ms between cilia and all cilia exhibited the same spatial asymmetry, as shown in Figure 5.1d. The resulting fluid flow has been measured using particle image velocimetry (PIV) and simulated using computational fluid dynamics. The results of these test for flaps with an inclination angle of 45° are depicted in Figure 5.2a.

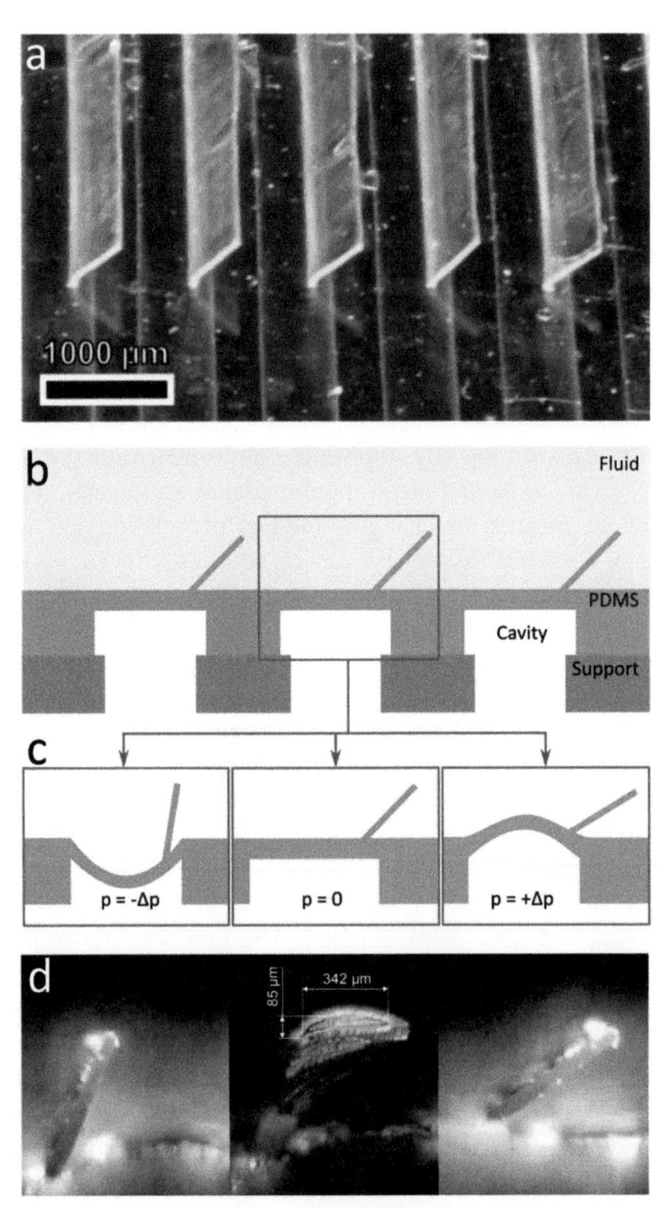

Figure 5.1 Schematic overview of base actuated pneumatic cilia that in essence are tilted flaps with an inclination angle of 45°, as shown in (a). These tilted flaps are located on top of a deformable PDMS membrane, as can be seen in (b). When actuating the cavity beneath the membrane, these plate-like cilia exhibit spatial asymmetry, as depicted schematically in (c) and using pictures in (d). Pictures adapted from [22] and [27].

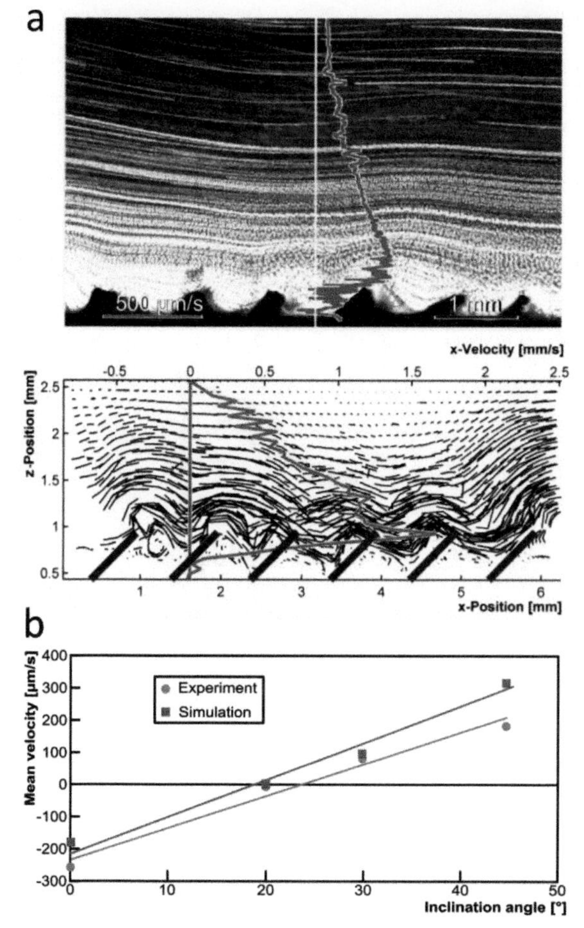

Figure 5.2 (a) Stream profile of the transported fluid for flaps with an inclination angle of 45°. Top: Experimentally obtained through PIV. Bottom: Simulated. (b) Comparison of mean velocities while influencing orientational asymmetry by changing flap inclination angle. Pictures adapted from [27].

These figures show a positive net velocity that is largest at the cilia tips. The influence of orientational asymmetry can be seen in Figure 5.2b, where it can be seen that increasing orientational asymmetry has a positive influence on net fluid flow, shifting it from an initial negative to a positive net fluid flow.

The influence of metachronal asymmetry has been experimentally studied, by changing the time shift between actuated flaps, while keeping temporal, orientational, and spatial asymmetry constant [24]. To observe

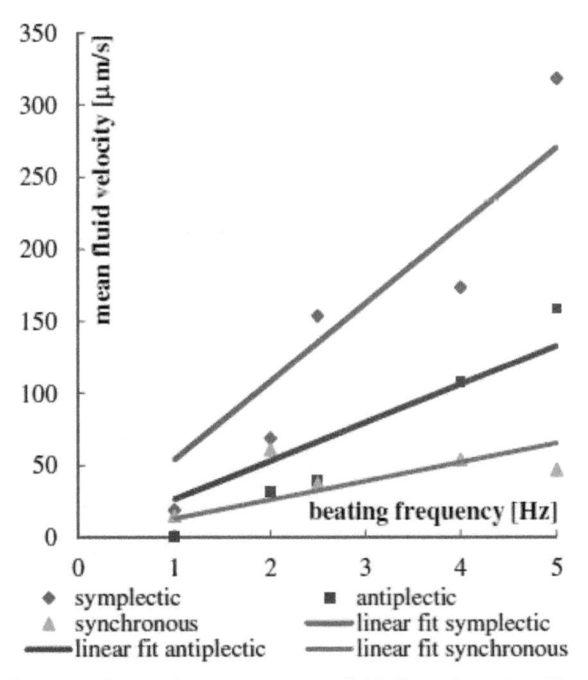

Figure 5.3 Influence of metachrony on mean fluid flow, for plate-like body actuated pneumatic cilia. Picture adapted from [27].

the fluid flow and evaluate the velocities, particle path-lines' pictures were acquired and used to calculate mean fluid velocities. The results of changing metachrony on this mean fluid flow can be seen in Figure 5.3, where beating frequencies were tested up to 5 Hz. As shown, synchronous beating leads to low mean velocities compared to when metachrony is introduced. This can be explained by the absence of a relative motion between the neighboring walls. This relative motion squeezes fluid out of the gaps between cilia into the upper layer, above the cilia tips, where it contributes to the main flow. It is also observed that symplectic metachrony (mean fluid velocity is in the same direction as the effective stroke) causes higher velocities than antiplectic metachrony. This is in contrast to the theoretical findings for low Reynold number ciliated fluid flow of Khaderi et al. [10], which showed for low Reynolds flow an increased performance of antiplectic metachrony over symplectic metachrony. As the base actuated pneumatic cilia operate at a Reynolds number of 5, it can be concluded that inertial effects play an important role on metachronally induced fluid flow [24].

5.3 Body Actuated Pneumatic Cilia

As shown above, base actuated pneumatic cilia can give clear insights in the influence of certain types of asymmetries on fluid flow. However, it is hard to distinguish the effect of the protruding cilia from the effect of the base itself. Further, the shape of the cilia remains straight throughout its stroke while cilia in nature often use a change in shape to propel fluid. This can be clearly seen in the beating pattern of natural cilia, as shown in Figure 5.4, where the cilia body is extended throughout the effective stroke and bended during the recovery stroke. Body actuated pneumatic cilia on the other hand try to mimic these advanced beating patterns by actuating the internal structure of the cilia itself rather than actuating the base structure. An artificial equivalent of this concept has been presented by Gorissen et al. [23], where a pneumatic bending microactuator is used as a biomimicking cilia. The artificial cilium consists of a PDMS cylinder with a length of 8 mm and a diameter of 1 mm, which has an inner cylindrical void with a diameter of 0.6 mm and an eccentricity with respect to the outer structure of 0.14 mm. Due to this eccentricity, the structure bends when the inner void is pressurized, generating the effective stroke, as can be seen in Figure 5.5a. When actuated, this cilium exhibits orientational asymmetry that can be adjusted by influencing the actuation frequency. This is shown in Figure 5.5b, where it is also clear that when the cilium is actuated above its natural bandwidth, it cannot complete its entire stroke. This stroke reduction causes the mean bending deformation to increase, resulting in an increased orientational asymmetry. Further, this pneumatic cilium can exhibit temporal asymmetry, by changing the rate of inflation and deflation, and when put in an array, metachrony is adjustable by actuating the cilia with a phase difference between them. These cilia thus lack only the ability of spatial asymmetry.

In a first test, a single cilium was immersed in an ink bath and actuated at 10 Hz, as shown in Figure 5.5a. The pneumatic input signal was provided using a fast switching pneumatic valve (Festo, MH2), which was also used to actuate the cilia array where each cilium is actuated using a separate valve. Consecutive images show high Reynolds fluid propulsion in the actuation direction, which is in accordance with the theoretical results of Khaderi et al. [2]. Subsequently, the net propulsion generated by the cilia array is measured in the configuration of no metachrony, symplectic metachrony, and antiplectic metachrony for duty cycles of 80% and 20% in a range of frequencies varying between 1 and 35 Hz. The results are summarized in Figures 5.6a and b. Over all frequencies, it is not possible to make concise conclusions on which type

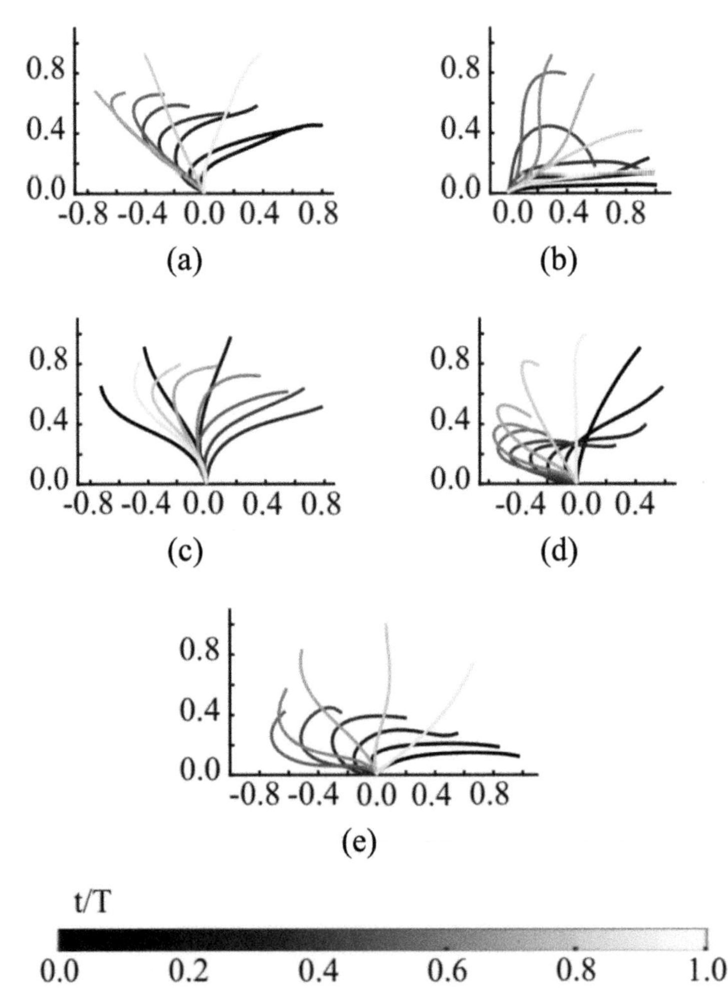

Figure 5.4 Beating patters of natural cilia for (a) *Paramecium*, (b) *Opalina*, (c) rabbit tracheal cilia, (d) *Sabellaria* gill, and (e) *Didinium*. Figure adapted from [26].

of metachrony is best or achieving fluid propulsion. This is in contrast to the conclusions with the base actuated pneumatic cilia. Below 11 Hz, in all cases, a flow directed in the opposite direction of the effective strokes of the cilia is observed. However, above 11 Hz, the change in the orientational asymmetry caused by the duty cycle at 80% appears to generate a flow reversal, even though the orientation angle does not shift in sign. PIV measurements, as shown in Figure 5.6c, explain these observations: generally, the single cilium

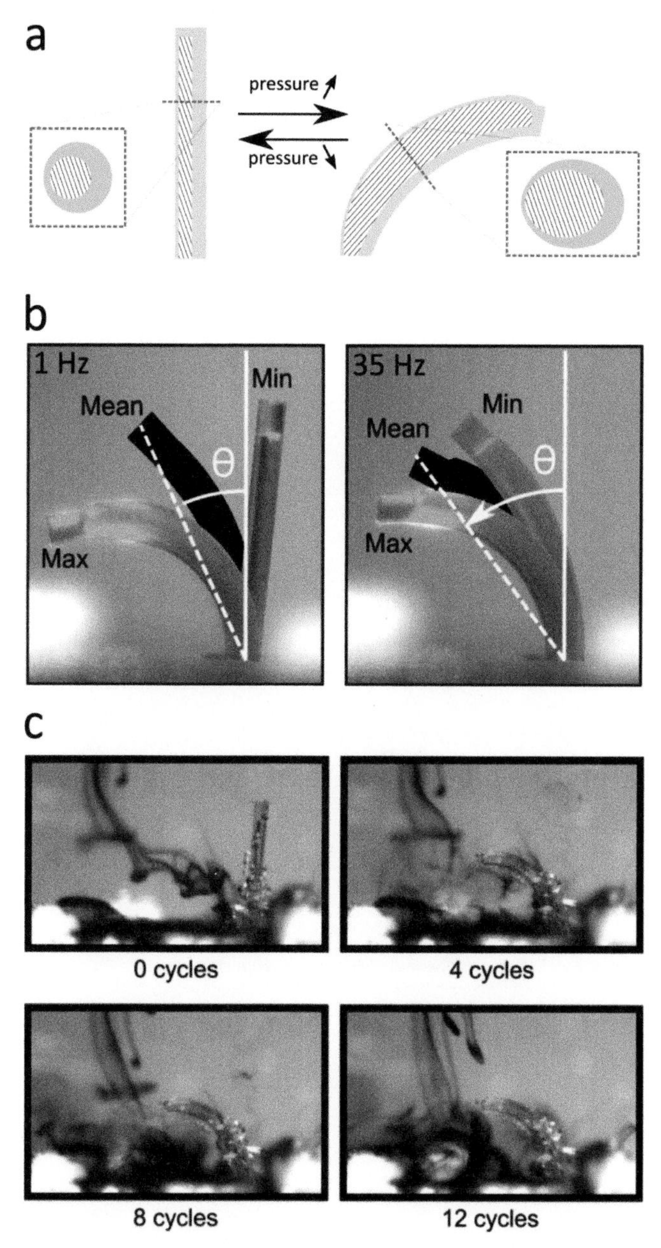

Figure 5.5 (a) Schematic overview of the pneumatic cilia without spatial asymmetry. (b) Stroke reduction that appears when the cilia is actuated at 35 Hz, compared to the full stroke at 1 Hz. As the mean bending angle changes with stroke reduction, the orientational asymmetry is also influenced. (c) High Reynolds fluid propulsion of a single cilium, showing a net fluid flow in the bending direction. Pictures adapted from [23].

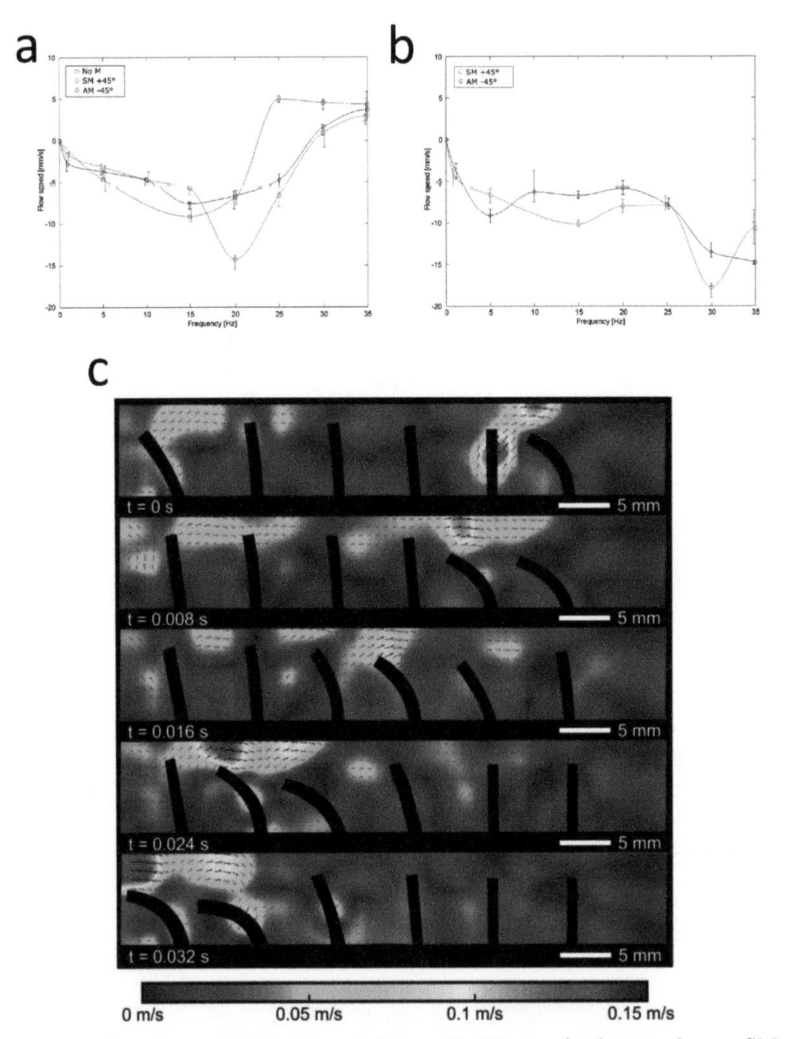

Figure 5.6 Effect of metachrony (no metachrony: No M, symplectic metachrony: SM, and anitplectic metachrony: AM) on flow speed, for an actuated cilia array with duty cycles of 80% (a) and 20% (b). (c) Velocity field for symplectic metachrony with a frequency of 20 Hz, a duty cycle of 20%, and a phase difference of 45°, registered through PIV. Pictures adapted from [23].

propels fluid in the same direction as its bending motion, but when it is placed in the array configuration, the momentum is blocked by the second proximal cilium which forces the fluid to go back over the actuated cilia and move in the opposite direction. This blockage is less pronounced when the duty cycle is 80% and the frequency is high, since in this case, the second cilium is in a more bended configuration, allowing the flow to move in the same direction as the bending direction. Finally, these tests show that for high Reynolds propulsion ($Re \approx 4000$), no spatial asymmetry is needed to achieve fluid propulsion.

5.4 Conclusion

During the previous decade, artificial cilia have been developed using a wide variety of actuation principles. Recently, new types of artificial cilia have been introduced that are using flexible pneumatic actuators to achieve propulsion, by deforming either the base of the cilia or the cilia itself. Compared to other actuation principles, these pneumatic cilia are uniquely suited to validate existing cilia propulsion theories, since these artificial cilia allow all types of asymmetries to be independently activated and adjusted in magnitude. This makes it possible to differentiate between asymmetry types and carefully analyze the contribution of each type to the overall fluid flow. Artificial pneumatic cilia are thus an important enabler to experimentally study these fascinating propulsion systems.

Acknowledgments

This book chapter has been made possible by the Fund for Scientific Research-Flanders (FWO) and the European Research Council (ERC starting grant HIENA).

References

[1] Purcell EM. Life at low reynolds-number. *American Journal of Physics*, 45(1):3–11, 1977.

[2] Khaderi SN, Baltussen MGHM, Anderson PD, et al. Breaking of symmetry in microfluidic propulsion driven by artificial cilia. *Physical Review*, 25;82(2):027302, 2010.

[3] Hussong J, Schorr N, Belardi J, et al. Experimental investigation of the flow induced by artificial cilia. *Lab on a Chip*, 11(12):2017–2022, 2011.

[4] Brennen C, and Winet H. Fluid-mechanics of propulsion by cilia and flagella. *Annual Review of Fluid Mechanics*, 9:339–398, 1977.

[5] Shields AR, Fiser BL, Evans BA, et al. Biomimetic cilia arrays generate simultaneous pumping and mixing regimes. *Proceedings of the National Academy of Sciences of the United States of America*, 107(36):15670–15675, 2010.

[6] Evans BA, Shields AR, Carroll RL, et al. Magnetically actuated nanorod arrays as biomimetic cilia. *Nano Letters*, 7(5):1428–1434, 2007.

[7] Fahrni F, Prins MWJ, and van Ijzendoorn LJ. Micro-fluidic actuation using magnetic artificial cilia. *Lab on a Chip*, 9(23):3413–3421, 2009.

[8] Vilfan M, Potocnik A, Kavcic B, et al. Self-assembled artificial cilia. *Proceedings of the National Academy of Sciences of the United States of America*, 107(5):1844–1847, 2010.

[9] Khaderi S, Hussong J, Westerweel J, et al. Fluid propulsion using magnetically-actuated artificial cilia – experiments and simulations. *Rsc Advances,* 3(31):12735–12742, 2013.

[10] Khaderi SN, Craus CB, Hussong J, et al. Magnetically-actuated artificial cilia for microfluidic propulsion. *Lab on a Chip*, 11(12):2002–2010, 2011.

[11] Wang Y, den Toonder J, Cardinaels R, et al. A continuous roll-pulling approach for the fabrication of magnetic artificial cilia with microfluidic pumping capability. *Lab on a Chip*, 16(12):2277–2286, 2016.

[12] Chen CY, Hsu CC, Mani K, et al. Hydrodynamic influences of artificial cilia beating behaviors on micromixing. *Chemical Engineering and Processing*, 99:33–40, 2016.

[13] Chen C-Y, Cheng L-Y, Hsu C-C, et al. Microscale flow propulsion through bioinspired and magnetically actuated artificial cilia. *Biomicrofluidics*, 9(3):034105, 2015.

[14] Chen CY, Lin CY, and Hu YT. Magnetically actuated artificial cilia for optimum mixing performance in microfluidics. *Lab on a Chip*, 13(14):2834–2839, 2013.

[15] den Toonder J, Bos F, Broer D, et al. Artificial cilia for active microfluidic mixing. *Lab on a Chip*, 8(4):533–541, 2008.

[16] Sareh S, Rossiter J, Conn A, et al. Swimming like algae: biomimetic soft artificial cilia. *Journal of the Royal Society Interface*, 10(78), 2013.

[17] van Oosten CL, Bastiaansen CWM, and Broer DJ. Printed artificial cilia from liquid-crystal network actuators modularly driven by light. *Nature Materials*, 8(8):677–682, 2009.

[18] Oh K, Smith B, Devasia S, et al. Characterization of mixing performance for bio-mimetic silicone cilia. *Microfluidics and Nanofluidics*, 9(4–5):645–655, 2010.

[19] Oh K, Chung J-H, Devasia S, et al. Bio-mimetic silicone cilia for microfluidic manipulation. *Lab on a Chip*, 9(11):1561–1566, 2009.
[20] Keissner A, and Bruecker C. Directional fluid transport along artificial ciliary surfaces with base-layer actuation of counter-rotating orbital beating patterns. *Soft Matter*, 8(19):5342–5349, 2012.
[21] Rockenbach A, and Schnakenberg U. The influence of flap inclination angle on fluid transport at ciliated walls. *Journal of Micromechanics and Microengineering*, 27(1):015007, 2017.
[22] Rockenbach A, and Schnakenberg U. Structured PDMS used as active element for a biomimetics inspired fluid transporter. *Lekar a technika - Clinician and Technology*, 2(45):37–41, 2015.
[23] Gorissen B, de Volder M, and Reynaerts D. Pneumatically-actuated artificial cilia array for biomimetic fluid propulsion. *Lab on a Chip*, 15(22):4348–4355, 2015.
[24] Rockenbach A, Mikulich V, Brucker C, et al. Fluid transport via pneumatically actuated waves on a ciliated wall. *Journal of Micromechanics and Microengineering*, 25(12):125009, 2015.
[25] Dauptain A, Favier J, and Bottaro A. Hydrodynamics of ciliary propulsion. *Journal of Fluids and Structures*, 24(8):1156–1165, 2008.
[26] Guo HL, and Kanso E. Evaluating efficiency and robustness in cilia design. *Physical Review*, 93(3):033119, 2016.
[27] Rockenbach AB. Fluidtransport an Grenzflächen durch pneumatisch angesteuerte strukturierte Oberflächen mit zilienähnlichen Strukturen, RWTH Aachen University, Aachen, 2017, PhD Thesis.

6

Blinking-Vortex Inspired Mixing with Cilia

Nathan Banka and Santosh Devasia

Mechanical Engineering Department, University of Washington,
Seattle, WA 98195-2600 USA
E-mail: nbanka@uw.edu; devasia@uw.edu

It is well known that symmetric reciprocal motions may not produce net flow at low Reynolds numbers [1, 2]. Therefore, there is a substantial interest in developing asymmetry, e.g., [3, 4]. For example, phase delays between adjacent cilia can be used to break the overall symmetry even if the motion of each cilium is symmetric and reciprocal [5]. Past works using numerical simulations [6, 7] and more recent experimental work [8] indicate that such non-simultaneous actuation improves pumping performance. Improved mixing performance was also demonstrated with cilia using geometric effects and sloshing-based actuation, e.g., in [9–11]. In biological systems, co-ordinated non-simultaneous actuation, i.e., metachronal waves, where the beating of the cilia forms a traveling wave is understood to result in fluid flow, e.g., [12]. Motivated by these results, experimental work in [13] demonstrated that using a blinking-vortex-inspired approach can break the motion symmetry and, thereby, enhance mixing. The simulation results using blinking-vortex theory and the experimental results are compared in this chapter. Details of the experimental system and the theoretical simulations are available in [13]. Moreover, models of the cilia, which can aid in optimizing the cilia design, are available in [14]. The experimental results closely follow the theoretical simulations, as shown below.

6.1 Cilia Actuation Scheme

The mixing strategy is based on the theory of the blinking vortex [15, 16]. The main idea is to use two spatially overlapping vortices that are alternately activated; this motivated the use of two cilia, each of which can generate a vortex. The cilia-based magnetic device capable of individual actuation is shown in Figure 6.1. Two cilia are individually controlled by point sources of magnetic field, which are positioned to the side of each cilia to cause deflection. Since the magnetic field falls off rapidly with distance, separable actuation can be achieved by allotting one magnetic field source to each cilium. For example, when a magnet is close to the slide on the left side, the left cilium deflects due to a distributed magnetic force f. Since the other cilium is far from the magnet, it is not affected by the magnet, and thus, it can be actuated independently.

The magnetic actuation method shown in Figure 6.1 was applied to a batch-mixing process on a standard 25 mm × 75 mm glass slide, as is commonly used for DNA microarray experiments [17–19]. The chamber was made from polydimethylsiloxane via a molding process and has interior dimensions of 40 mm × 18 mm × 3.8 mm as described in [13]. The cilia presented here have lengths of 13 mm, which is similar to recent work [8] that demonstrated effective pumping using pneumatically actuated artificial cilia that were 8 mm long and 1 mm in diameter, and used a similar ratio of cilium length to channel width.

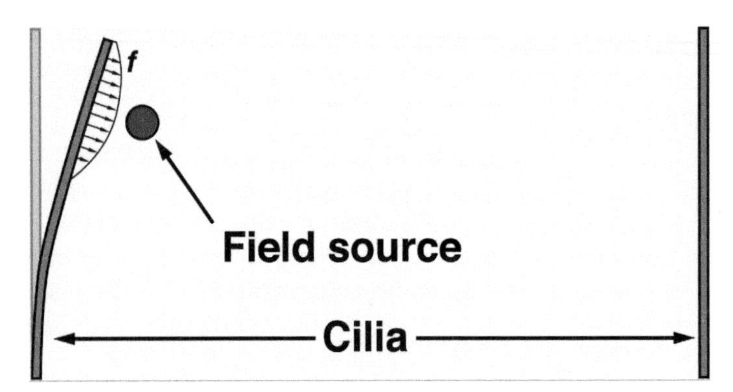

Figure 6.1 Schematic of individual control of cilia. Figure is from [13].

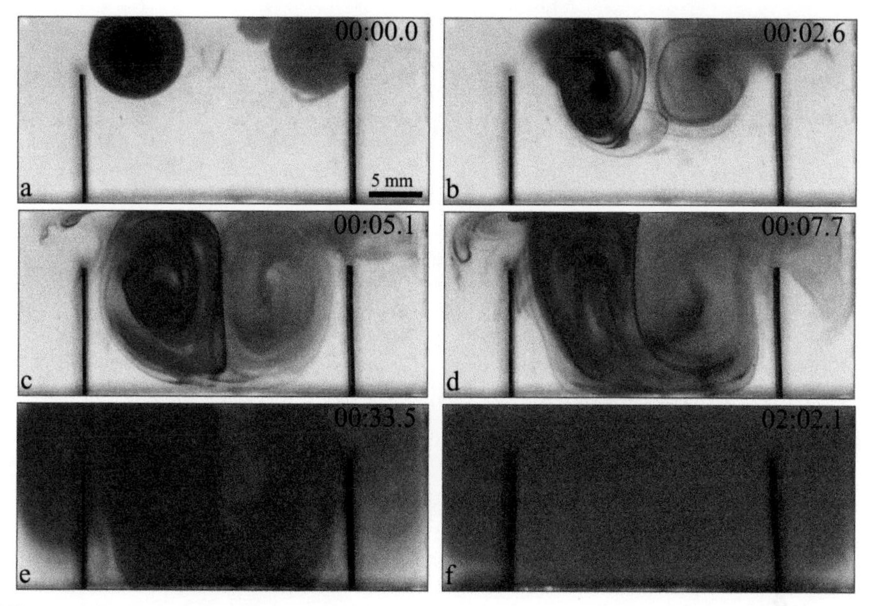

Figure 6.2 Experimental images with the simultaneous beat pattern. Figure is from [13]. Timestamps (mm:ss) are shown in the top corner of each image. (a–d) Initial distribution of ink and result after each of the first three cycles ($\frac{t}{T} = 0, 2, 4, 6$), where t is the time and the time period between the successive cilia beats for the simultaneous beat pattern is $2T$. A boundary is formed between the right and left sides due to symmetry (c); however, unsteady processes sometimes break symmetry and promote mixing (d). (e) As the experiment continues, the boundary remains and mixing largely occurs when ink moves from one side to the other along the bottom boundary. (f) Image with $c_v(b^*) = 0.05$.

6.2 Experiments and Simulations

The cilia-actuation patterns, modeled after the blinking vortex [15], are discrete sequences. In particular, the actuation aims to beat the left cilium n times (for n time steps), beat the right cilium n times, and then repeat. This pattern is denoted Ln-Rn and is compared with the simultaneous actuation case (referred to as "Simul"), where both cilia beat together at odd-numbered time-steps, and neither cilium beat on even-numbered time-steps.

The blinking vortex theory was simulated for a rectangular domain (to match the experimental work). Note that the original blinking vortex theory [15] was developed to understand chaotic behavior of iterated alternating mappings in inviscid, incompressible, two-dimensional flow. In its original development, the fluid domain is taken to be a circular region with radius a. An ideal point vortex with circulation was introduced into this domain whose

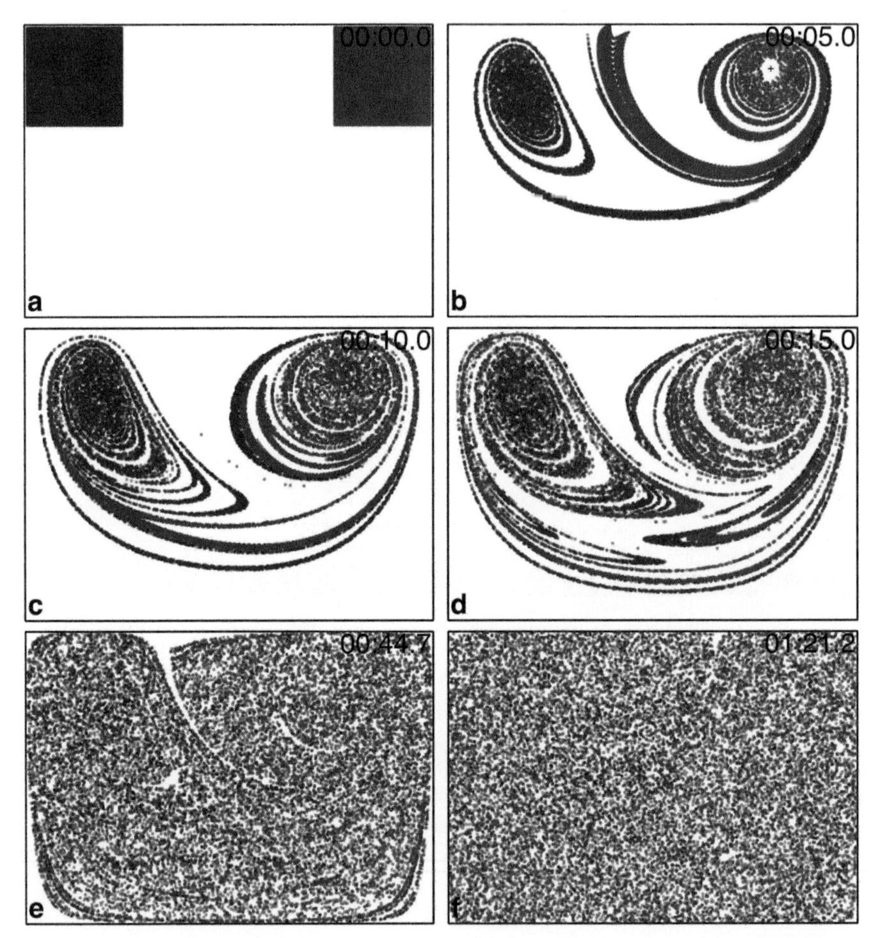

Figure 6.3 Simulations results for the simultaneous beat pattern for the experimental results in Figure 6.2. Timestamps (mm:ss), shown in the top corner of each image, are the same as in the experimental images in Figure 6.2. (a–d) Initial distribution of ink and result after each of the first three cycles ($\frac{t}{T} = 0, 2, 4, 6$). A boundary is formed between the right and left sides due to symmetry (c). (e) As the experiment continues, the boundary remains and mixing does not occur in simulations as seen in (f). Copyright: N. Banka.

position alternates between two points. Although the flow for each half-cycle is non-mixing, alternating between leads to chaotic behavior. It was shown that the chaotic or non-chaotic nature of the flow is determined by the vortex position and alternation period. Details of the simulations for applying the blinking-vortex theory for the rectangular domain are described in [13].

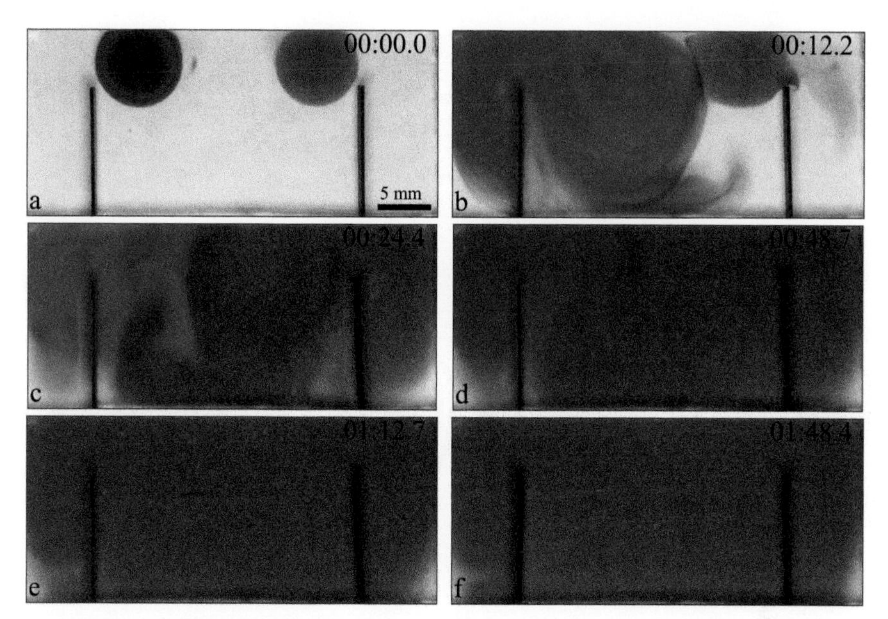

Figure 6.4 Experimental images with the L10-R10 beat pattern. Figure is from [13]. Timestamps (mm:ss) are shown in the top corner of each image. (a) Initial distribution of ink. (b) After half a cycle ($\frac{t}{T}$ = 10), a uniform blue region is found in the region of influence of the left cilium. Note that the time period between successive cilia beats during this half cycle is T. (c) After a full cycle ($\frac{t}{T}$ = 20), the action of the right cilium has made the right half uniform, drawing in some of the blue ink. (d–e) After two further cycles ($\frac{t}{T}$ = 40, 60), it is observed that the boundary between colors is consistent after each full cycle. However, the color of the two uniform regions converges to a uniform value after each cycle. (f) After sufficient cycles, the color reaches the threshold, $c_v(b^*) = 0.05$.

6.3 Results

The simultaneous actuation mode does not lead to mixing as seen in the experimental results shown in Figure 6.2. Ideally, the flow would be laminar and the device would be perfectly symmetric. This implies that mixing should occur only by diffusion along the line of symmetry, which would be slow. This is confirmed by the simulations shown in Figure 6.3. If the flow were laminar and the device were perfectly symmetric, the expected outcome would be that mixing would occur only by diffusion along the line of symmetry. However, due to unsteady flow, ink is also transferred from one side to the other, primarily along the bottom of the image, e.g., as seen in Figure 6.2d, and some amount of mixing is seen. The relative standard

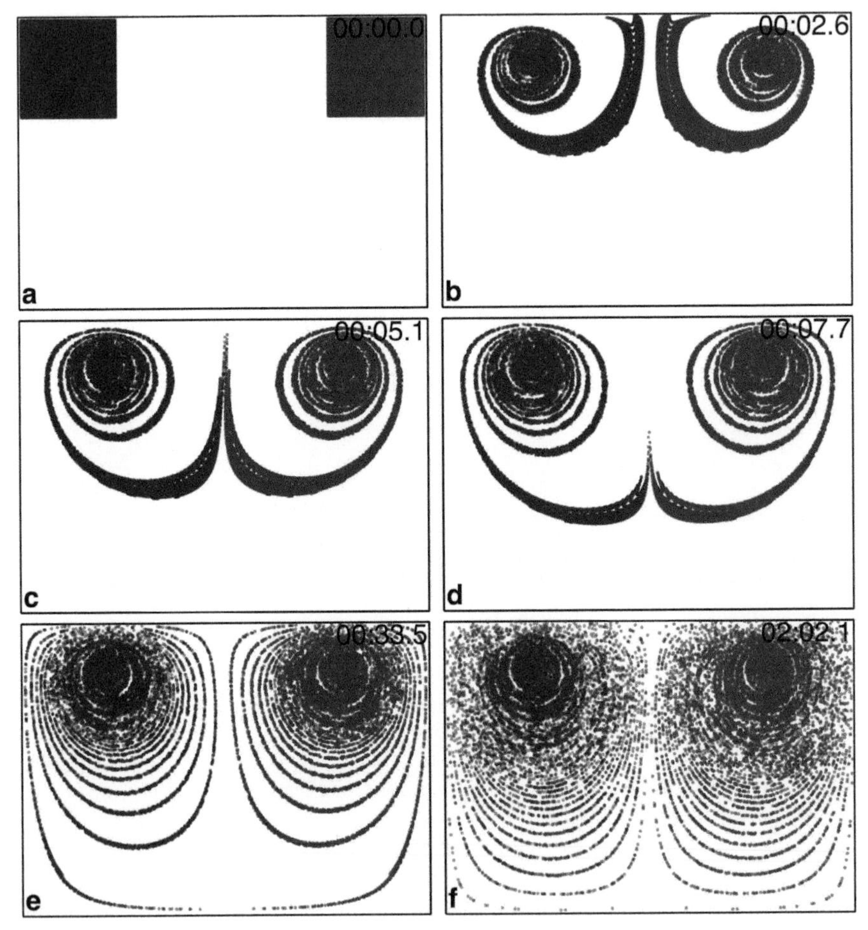

Figure 6.5 Simulations results for the blinking-vortex simulations corresponding to the L10-R10 beat pattern. Timestamps (mm:ss), shown in the top corner of each image, are the same as in the experimental images in Figure 6.4. (a–e) The mixing from the blinking vortex simulations is close to the experimental results. (f) After a similar number of cycles, as in the experiments, good mixing is seen in the simulations. Copyright: N. Banka.

deviation, denoted $c_v(b^*)$, was used to measure the variation of the blue-yellow b^* channel and hence the overall color variation in the chamber. This measure, of the progress of the mixing, is also called the coefficient of variation (hence the symbol c_v), and is a standard metric in mixing [20, 21].

Individual actuation promotes mixing. For example, as the left cilium beats, circulation is generated in the nearby fluid. After many beats, a uniform

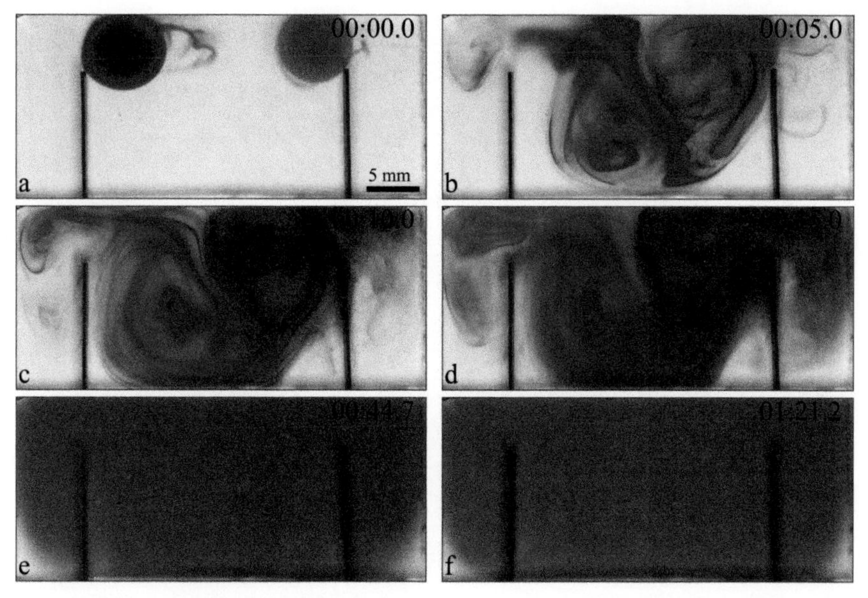

Figure 6.6 Experimental video images with the L2-R2 beat pattern. Figure is from [13]. Timestamps (mm:ss) are shown in the top corner of each image. (a–d) Images showing the initial development of the flow after each of the first three cycles ($\frac{t}{T} = 0, 4, 8, 12$). (e) Image showing the progression of the mixing flow. (f) Image with $c_v(b^*) = 0.05$.

blue region is seen on the left side. Note that the blue region on the left side covers more than half of the chamber and extends into the area that will be circulated by the right cilium, e.g., see Figure 6.4b. When the actuation pattern switches, the right cilium will cause the fluid on the right side of the chamber to circulate. If the actuation of the left cilium introduced blue fluid into the region of circulation on the right-hand side, then the result after many beats will be a uniform region whose color is a red-purple as seen in Figure 6.4c. When the cycles are repeated, whenever the actuation pattern switches, the part of the chamber circulated by the newly active cilium will be the less uniform side of the chamber. Eventually, both halves of the chamber converge toward a mean color, as seen in Figure 6.4d–f. Note that the simulation results seen in Figure 6.5 show that the mixing with the blinking-vortex follows the mixing seen in the experimental results in Figure 6.4.

A practical actuation pattern Ln-Rn, with finite n, will not necessarily achieve the full uniformity shown in Figure 6.4. Moreover, as one cilium continues to beat, it tends to make the fluid on its side more uniform, which

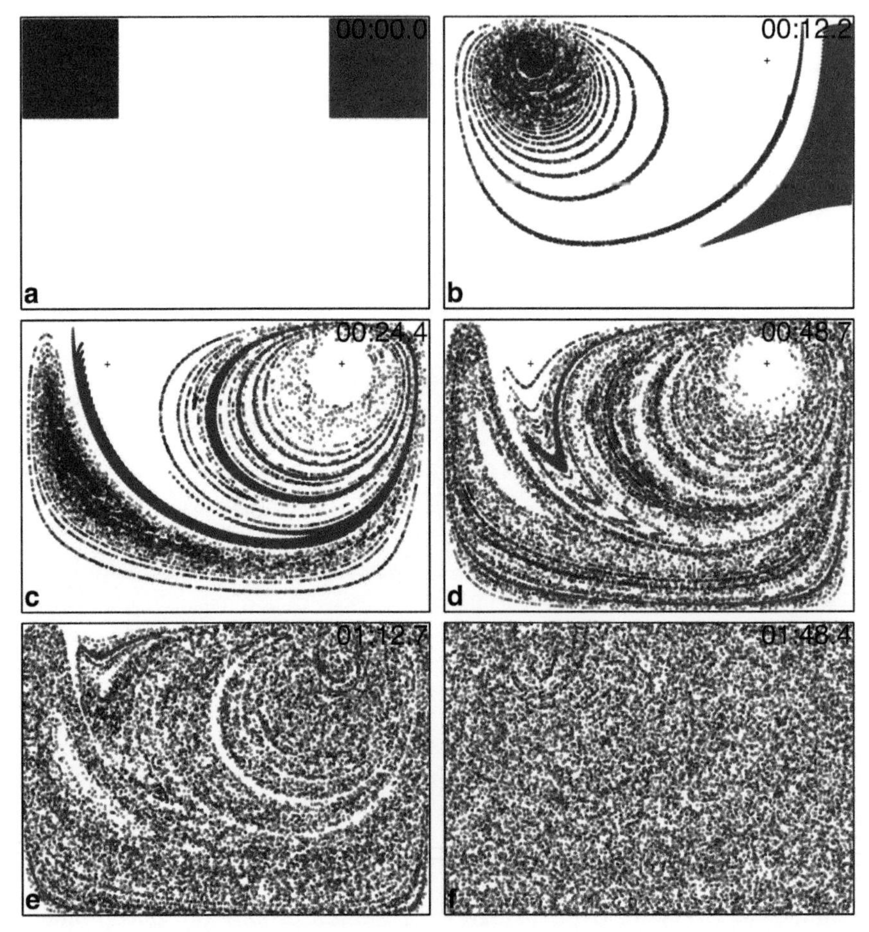

Figure 6.7 Simulations results for the blinking-vortex simulations corresponding to the L2–R2 optimal beat pattern. Timestamps (mm:ss), shown in the top corner of each image, are the same as in the experimental images in Figure 6.6. (a–e) The mixing from the blinking vortex simulations is close to the experimental results. (f) After a similar number of cycles, as in the experiments, good mixing is seen in the simulations. Copyright: N. Banka.

results in diminishing returns as the number of beats per cycle, n, increases. Note that the circulation of the fluid in the chamber is most effective at mixing the colors just after the actuation switches, when the fluid on the active side of the chamber is least uniform. At the same time, several beats are needed for the vortex to reach into the other half of the chamber, as shown in Figure 6.4b. Due to these competing effects, one expects an optimal number of beats n in the Ln-Rn patterns to balance the diminishing returns when continuing the beats against the time needed for the vortex to fully develop.

In general, the optimal number of beats can be expected to depend on system parameters such as the magnet placement and fluid properties. For the specific experiments, it was found both by simulations and experiments that the L2-R2 pattern mixes the fastest. The experimental results are presented in Figure 6.6, which match the trends seen in the simulation results in Figure 6.7. Compared to the simultaneous pattern, the L2-R2 pattern reaches a mixed state (corresponding to relative standard deviation $c_v(b^*) < 0.05$) in 38% less time in experiments. The tradeoffs between small n and large n in the Ln-Rn pattern, discussed above, can be seen by comparing images of the flow from the L2-R2 pattern in Figure 6.6 and the L10-R10 pattern in Figure 6.4.

The above results show that blinking-vortex type mixing could be achieved using artificial cilia. Simulation results show similar trends to the experimental results, which can aid in optimizing the actuation patterns. This indicates that such simulations could be used in future works to optimize the size, placement, and number of cilia in the mixing chamber for improved mixing.

References

[1] Purcell EM. Life at low Reynolds number. *Am. J. Phys.*, 45(1):3–11, 1977.

[2] Kim YW, and Netz RR. Pumping fluids with periodically beating grafted elastic filaments. *Physical Review Letters*, 96(15):158101, 2006.

[3] Khaderi S, Baltussen M, Anderson P, *et al.* Breaking of symmetry in microfluidic propulsion driven by artificial cilia. *Physical Review E*, 82(2):027302, 2010.

[4] Downton M and Stark H. Beating kinematics of magnetically actuated cilia. *EPL (Europhysics Letters)*, 85(4):44002, 2009.

[5] Khaderi S, den Toonder J, and Onck P. Fluid flow due to collective non-reciprocal motion of symmetrically-beating artificial cilia. *Biomicrofluidics*, 6(1):014106, 2012.

[6] Khaderi S, Craus C, Hussong J, *et al.* Magnetically-actuated artificial cilia for microfluidic propulsion. *Lab on a Chip*, 11(12):2002–2010, 2011.

[7] Gauger EM, Downton MT, and Stark H. Fluid transport at low Reynolds number with magnetically actuated artificial cilia. *The European Physical Journal E*, 28(2):231–242, 2009.

[8] Gorissen B, de Volder M, and Reynaerts D. Pneumatically-actuated artificial cilia array for biomimetic fluid propulsion. *Lab on a Chip*, 15(22):4348–4355, 2015.

[9] Oh K, Chung J, Devasia S, *et al.* Bio-mimetic silicone cilia for microfluidic manipulation. *Lab on a Chip by the Royal Society of Chemistry*, 9(11):1561–1566, 2009.

[10] Kongthon J, Mckay B, Iamratanakul D, *et al.* Added-mass effect in modeling of cilia-based devices for microfluidic systems. *Journal of Vibration and Acoustics*, 132(2):024501, 2010.

[11] Kongthon J, Chung J, Riley JJ, *et al.* Dynamics of cilia-based microfluidic devices. *Journal of Dynamic Systems, Measurement and Control*, 133(5):051012, 2011.

[12] Brennen C, and Winet H. Fluid mechanics of propulsion by cilia and flagella. *Annual Review of Fluid Mechanics*, 9(1):339–398, 1977.

[13] Banka N, Ng YL, and Devasia S. Individually controllable magnetic cilia: mixing application. *Journal of Medical Devices - Transactions of the ASME*, 11(3):031003, 2017.

[14] Banka N, and Devasia S. Nonlinear models for magnet placement in individually actuated magnetic cilia devices. *ASME Journal of Dynamic Systems, Measurement, and Control*, 140(6):061011, 2017.

[15] Aref H. Stirring by chaotic advection. *Journal of Fluid Mechanics*, 143:1–21, 1984.

[16] Wiggins S, and Ottino JM. Foundations of chaotic mixing. *Philosophical Transactions of the Royal Society of London A: Mathematical, Physical and Engineering Sciences*, 362(1818):937–970, 2004.

[17] Dugas V, Broutin J, and Souteyrand E, Mixing and dispensing homogeneous compounds of a reactant on a surface. *US Patent 7,772,010.*

[18] Wei C-W, Cheng J-Y, Huang C-T, *et al.* Using a microfluidic device for 11 DNA microarray hybridization in 500 s. *Nucleic Acids Research*, 33(8):e78–e78, 2005.

[19] Microarrays: Table of suppliers. *Nature*, 442(7106):1071–1072, 2006. [Online]. *Available at*: http://dx.doi.org/10.1038/4421071a

[20] Paul EL, Atiemo-Obeng VA, and Kresta SM, *Handbook of Industrial Mixing: Science and Practice*. John Wiley & Sons, 2004.

[21] McQuain MK, Seale K, Peek J, *et al.* Chaotic mixer improves microarray hybridization. *Analytical Biochemistry*, 325(2):215–226, 2004.

7

Magnetic Thin-film Cilia for Microfluidic Applications

Srinivas Hanasoge, Peter J. Hesketh and Alexander Alexeev

George W. Woodruff School of Mechanical Engineering,
Georgia Institute of Technology, Atlanta, GA, USA, 30332
E-mail: srinivasgh@gatech.edu; peter.hesketh@me.gatech.edu;
alexander.alexeev@me.gatech.edu

Biological organisms use hair-like cilia and flagella to perform fluid manipulations essential for their function and survival. We describe the fabrication, kinematics, and applications of biomimetic magnetic cilia that can be integrated into electromechanical devices to enhance their functionality. The magnetic cilia are fabricated using micromachining and are actuated with a rotating permanent magnet. The cilia beat in a time-irreversible fashion enabling fluid transport at a low Reynolds number. We demonstrate applications of the cilia for creating metachronal waves, fluid mixing, and bacteria capture.

7.1 Introduction

Dr. Feynman has remarked that "there is plenty of room at the bottom" pointing to the potential of developing micro and nano-devices. Today, the use of semiconductor fabrication technology for making microscale electromechanical devices (MEMS) has become widespread and such devices are often used as sensors and actuators in various applications. Biomimetic MEMS devices have gained significant attention in the recent past. Researchers have developed devices for performing microscale sensing, actuation, and other important functions in a manner that is inspired by biological organisms. In this context, microscale fluid handling is a particularly promising field for biomimetic MEMS design, as all of the biology exists in or around fluids. Such biomimetic fluid manipulation can be extremely useful in various lab-on-a-chip platforms for biomedical applications [1].

Achieving fluid transport at the micrometer length scales is difficult owing to the lack of inertial effects [2, 3]. A mere symmetrically reciprocating motion is not sufficient to produce any net fluid transport [3]. Microorganisms and cells that usually operate in these inertia-less regimes have evolved to use cilia and flagella with complex beating patterns yielding a net fluid transport [4–6]. Such complex beating cilia and flagella are imperative to the organism in performing vital biophysical functions.

The development of synthetic cilia that can perform similar non-reciprocal, spatially asymmetric beating pattern could prove useful for fluid and particle transport in lab-on-a-chip and bio-MEMS devices [7–11]. Several kinds of artificial systems have been experimentally demonstrated using electrostatic [12], pneumatic [13], and chemical actuation [14]. Among different approaches, magnetic actuation [15–20] is promising due to the relatively simple realization and operation. In this case, magnetic cilia can be actuated from a distance with no interference of the magnetic field with biological samples. Magnetically actuated ciliary systems typically comprise a magnetic structure that can deform elastically and an external magnetic field that drives the magnetic structure. Researchers have demonstrated cilia made of chains of magnetic beads [21], magnetic particles in a polymer [16, 22], and other similar materials [23]. Permanent magnets and electromagnet setups were used to create alternating magnetic fields for synchronous actuation of ciliary arrays [16, 21, 24, 25].

In what follows, we discuss a surface micromachining technique for fabricating microscale magnetic cilia and study their operating mechanism in creating fluid transport when actuated by a rotating permanent magnet. The fabrication process results in soft magnetic filaments made of nickel–iron alloy which has a high magnetic susceptibility, ensuring high magnetic forcing. Furthermore, the fabrication process is simple and employs standardized protocols of lithography and metal deposition, and relies on a two-step lithographic process with a quick processing time for making artificial cilia. The process yields highly reproducible cilia of various shapes and sizes.

7.2 Fabrication of Magnetic Cilia

We follow a simple UV lithographic process to fabricate the artificial cilia [20, 26]. The various steps of the fabrication are indicated in a flowchart

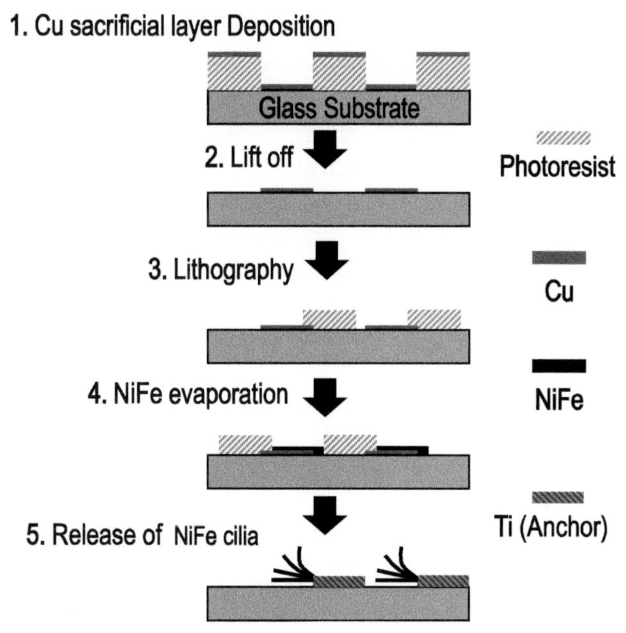

Figure 7.1 Flowchart of fabrication process involving surface micromachining of nickel-iron alloy on a glass substrate. The fabrication involves depositing Cu sacrificial layer first, followed by a layer of NiFe. A titanium anchor is deposited to attach on the edge of the thin filament to the substrate. Removing the Cu sacrificial layer leaves a free-standing elastic NiFe filament. Adapted from [20].

presented in Figure 7.1. The fabrication is performed on a soda-lime glass slide which is cheap and easily available. A clean glass slide is first spin coated with negative photoresist (NR9 1500PY, Futurrex, Inc., USA) and cilia features are imprinted using UV lithography. Developing the features in RD6 developer (Futurrex, Inc., USA) removes the resist and exposes the glass. Then a copper sacrificial layer of thickness 50 nm is deposited. Followed by this, a nickel–iron (80:20 Ni:Fe permalloy) layer of thickness 70 nm is deposited. We use an RF sputtering system for the metal deposition (UniFilm Technologies, USA). Lift-off is then performed to remove the excess resist and metal. This leaves the cilia features on the surface.

The next step is to deposit an anchor for holding the cilia permanently attached to the substrate. The second lithographic step is performed to obtain anchors which are deposited with 150 nm of titanium. The anchor layer sticks to the glass and ensures the NiFe cilia are firmly held on the substrate. The cilia are released by removing the sacrificial copper layer by dissolving it in

5% ammonium hydroxide which selectively etches the copper. The stresses due to sputtering NiFe alloy curl the film away from the substrate when the supporting layer is removed. This leaves a free-standing elastic filament of NiFe film that can be actuated by an external magnetic field. This process is robust and highly reproducible. Moreover, the fabricated devices can be stored for extended periods of time, and the sacrificial layer can be removed immediately before the experiment. The dimensions and geometry of the cilia can be readily modified by changing the lithographic mask and NiFe film deposition conditions.

An array of cilia fabricated using this method can be actuated subjected to a uniform rotating magnetic field using a rotating permanent magnet (12 mm diameter, K&J magnetics-D8X0DIA, USA). The large size of the magnet compared to the cilia ensures a uniform magnetic field actuating cilium motion. A schematic of our experimental setup is shown in Figure 7.2a. As the magnet rotates, the magnetic cilia orient themselves in the direction of the magnetic field leading to their cyclic oscillations. In Figure 7.2b–e, we show a ciliary array at different rotational positions of the actuating permanent magnet. These images are recorded from beneath the substrate as indicated in Figure 7.2a and show that all the cilia oscillate synchronously throughout the beating pattern.

7.3 Fluid Transport by An Array of Beating Magnetic Cilia

To probe whether the oscillating magnetic cilia can produce a net fluid flow, we conduct experiments in which flow patterns around beating cilia are visualized using fluorescent microparticles. We seed fluid with micrometer-sized fluorescent particles (1 μm FluoSpheres^{TM} – F13080, Thermofisher Scientific, USA) and register their motion using a standard epifluroscence microscope (Nikon – Eclipse Ti) while the array of cilia is subject to a rotating magnetic field. Figure 7.3 shows a snapshot of a fluid flow pattern produced by a ciliary array, where the trajectories of the fluorescent particles indicate fluid transport. In this experiment, cilia are arranged in four columns with four rows of cilia. Each row contains 15 individual cilia. The cilia in the successive columns are anchored at opposite edges. This results in cilia facing opposite directions in the four columns, as indicated by the arrows. The columns of cilia pump fluid in the direction of the arrows, resulting in a circulatory flow pattern shown by the streaklines generated by the fluorescent particles.

The dimensions of each individual cilium in the array are 200 μm $\times$ 20 μm $\times$ 70 nm, and the actuation frequency in this experiment is 50 Hz.

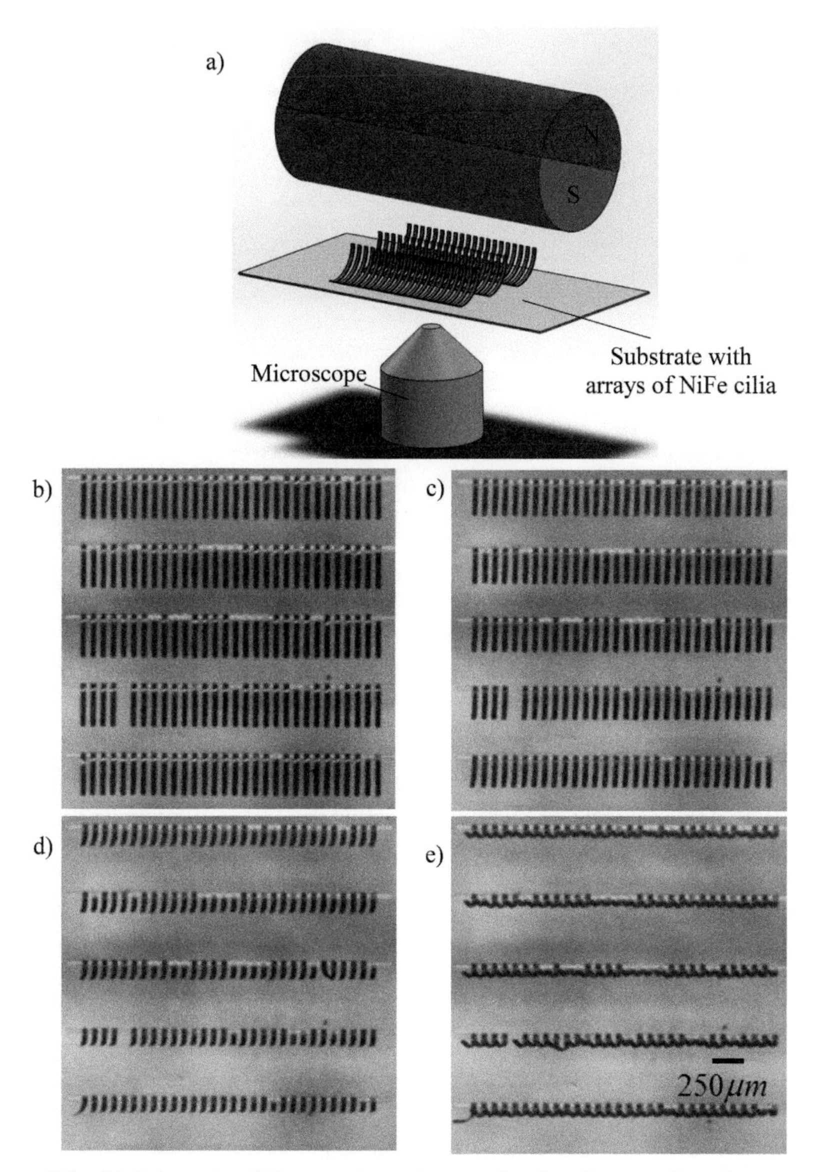

Figure 7.2 (a) Schematic of the experimental setup showing the actuation of an array of cilia using a permanent rotating magnet. (b–e) Series of experimental images of an array of cilia actuated by a rotating magnet, as viewed from the bottom. (b) Position of cilia when the magnetic field is horizontal and cilia are relaxed. (c and d) Changing positions of the cilia as the magnetic field is being rotated from the horizontal to the vertical direction. (e) Position of cilia when the field is normal to the substrate. Copyright S. Hanasoge, PJ. Hesketh, A. Alexeev, 2018.

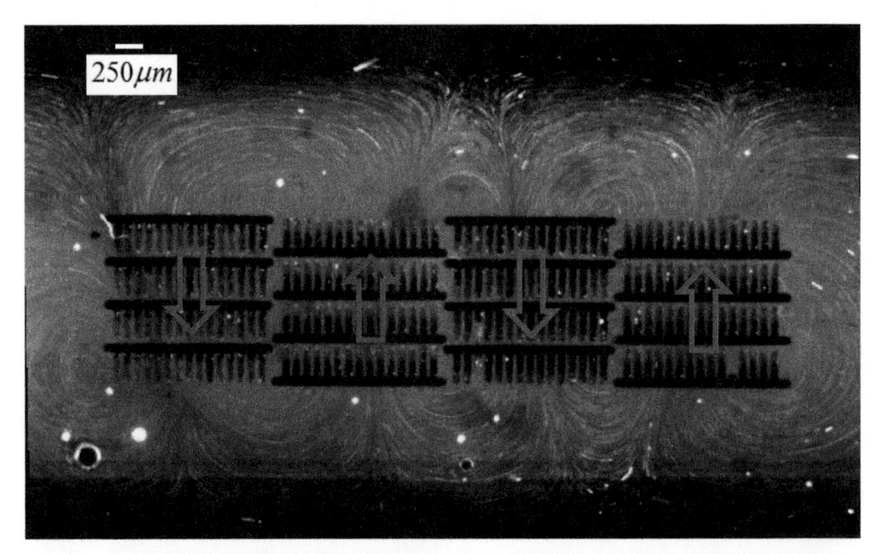

Figure 7.3 Fluid circulation produced by an array of magnetic cilia. The four columns of cilia are anchored on the opposite ends producing fluid pumping in the directions shown by the arrows. This leads to the fluid circulations as shown by the streaklines. Cilium dimensions are 200 μm $\times$ 20 μm $\times$ 70 nm and the actuation frequency is 50 Hz. Copyright S. Hanasoge, PJ. Hesketh, A. Alexeev, 2018.

These experimental conditions lead to a Reynolds number, $Re = \rho LWf/\mu \approx 0.1$, that represents the ratio of inertial forces to viscous forces acting on the cilia. Here, ρ is the fluid density, L is the length, W is the width, f is the frequency, and μ is the dynamic viscosity. For such values of Reynolds number, inertial effects are negligible and a time-irreversible motion with a spatial asymmetry of the beating pattern is required to create a net fluid transport [15]. Thus, the result in Figure 7.3 indicates that magnetic cilia are able to generate a time-irreversible motion leading to a net fluid flow. In the next section, we explore the beating kinematics of individual magnetic cilium to understand the mechanism leading to fluid transport.

7.4 Kinematics of Magnetic Cilia

The ability of an array of magnetic cilia to produce fluid circulation has to be related to time-irreversible, spatially asymmetric oscillations. To investigate beating kinematics, we visualize the stroke pattern of an individual isolated cilium under periodic magnetic forcing. An experimental setup for imaging the cilium beating pattern is shown in Figure 7.4. Here, the glass substrate is placed vertically such that the plane in which the cilium oscillates can be recorded using a microscope with a CCD camera attached.

When a soft magnetic thin filament is subjected to a uniform external magnetic field, it gets magnetized. If the filament has high magnetic permeability and a high aspect ratio, the local magnetization can be assumed to be tangential to the filament length. For such a scenario, the local magnetic couple acting at each point along the length of the filament is proportional to $\sin(2\theta)$, where θ is the angle between magnetic field **B** and the local filament axis [12, 17, 18, 20, 27]. For example, if a filament makes an angle of 45° with respect to the external magnetic field, it experiences the maximum magnetic moment. On the other hand, when the filament is aligned with the magnetic field, the magnitude of the magnetic moment is negligibly small. Thus, the magnetic moment acts to re-orient the filament along the direction

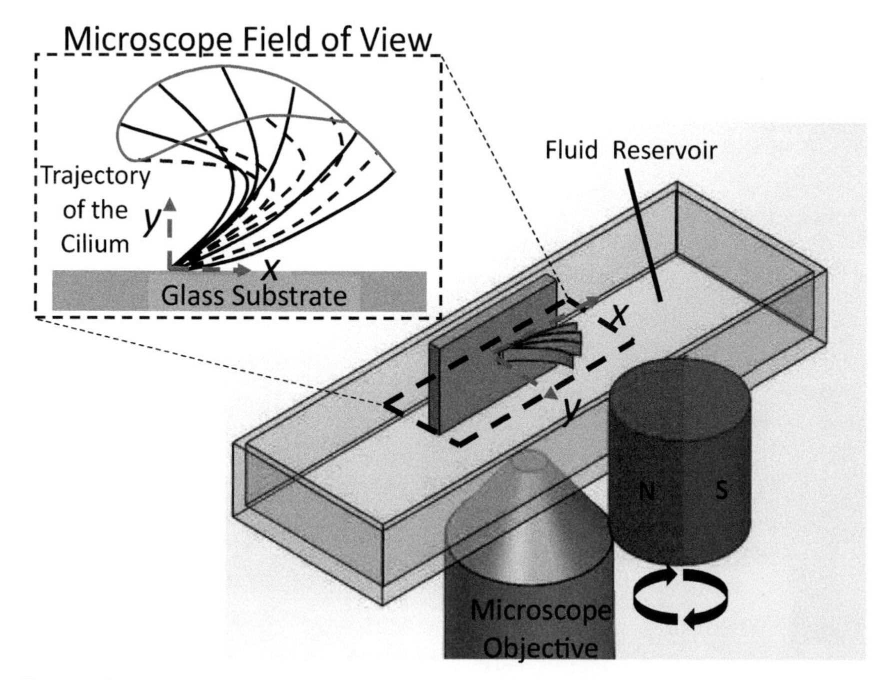

Figure 7.4 Experimental setup to capture the side-view images of the cilium beating. The substrate is placed on its vertical such that the plane of cilium beating is aligned with the microscope plane of view. The inset shows the expected bending pattern in the cilium. Adapted from [20].

of the external field. A similar magnetic moment arises in the case of thin film cilia with one edge anchored to the substrate and magnetized by an external magnetic field. The local magnetic moment proportional to $\sin(2\theta)$ acts on the anchored elastic cilia causing their bending. As a result, when the direction of the external magnetic field changes, the cilia bend accordingly to orient along the direction of the field [12, 17, 18, 20, 27].

In our experiments, magnetic field is rotated in the *xy* plane with its axis along the vertical direction as shown in Figure 7.4. As a result, the cilium beats in the plane of microscope view as the magnet rotates (see the inset of Figure 7.4). In this configuration, the magnetic field can be rotated either counter-clockwise (CCW) or clockwise (CW) direction. The CCW and CW rotating magnetic fields lead to a significantly different beating pattern of a cilium [19] which we discuss in the following sections.

7.4.1 Counter-Clockwise Rotation of Magnetic Field

When the magnet is rotated in the CCW direction, the cilium produces a large amplitude, spatial asymmetrical beating. An example of such a motion is shown in Figure 7.5 as a series of overlapped images of a cilium actuated by a CCW rotating magnet. The cilium tip follows a closed trajectory shown in blue during the forward stroke and in red during the recovery stroke. Differences in cilium motion during the forward and recovery strokes lead to a high degree of spatial asymmetry of cilium beating pattern required to create a local fluid transport [19].

Computer simulations are instrumental in gaining important insights into the behavior of magnetic cilia beating in a viscous fluid [12, 17, 18, 20, 27]. Here, we employ a 2D fluid–structure model using an arbitrary Lagrangian–Eulerian method in COMSOL. The magnitude of the moment is set to match the cilium deflection in the experiment. The fluid far from the cilium is assumed to be stationary. Simulation results are presented in Figure 7.5b and show close agreement with the results obtained in the experiments for the CCW rotation of magnetic field (Figure 7.5a).

Figure 7.6 shows selected positions (*a–e*) of a cilium during an oscillation cycle obtained using the computational model. The directions of the magnetic field corresponding to these positions are shown by the black arrow. The cilium starts the forward stroke at position *a*, which is the rightmost position in the cycle. In this position, most of the cilium is aligned with the direction of the magnetic field except a relatively short curved section near the base. Thus, at this position, the local magnetic moment, which is indicated by the

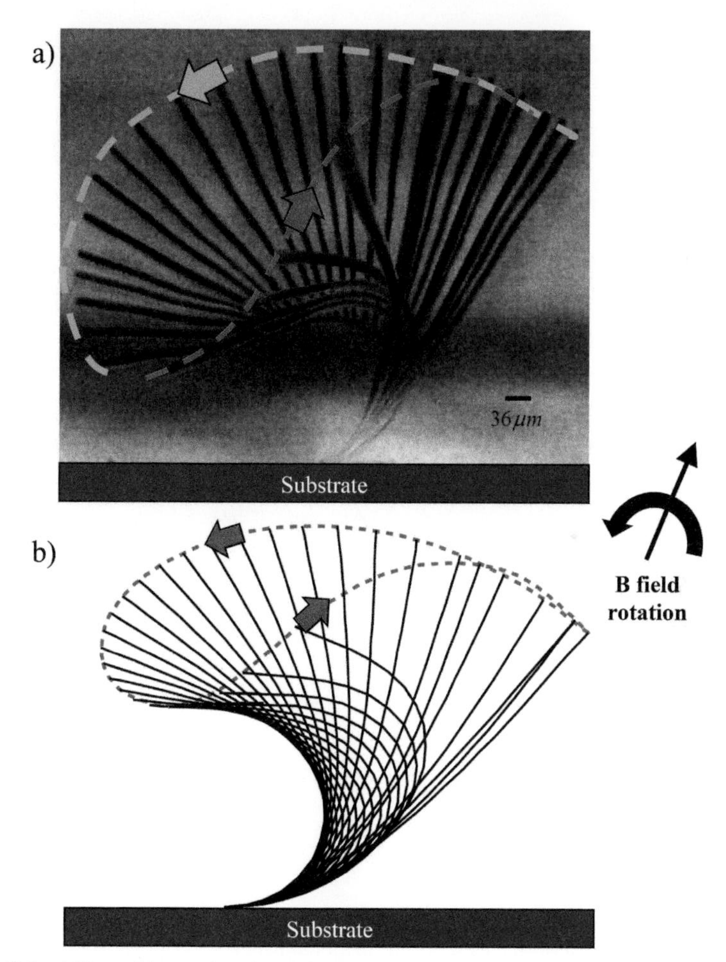

Figure 7.5 Cilium kinematics for CCW rotation of the magnetic field. (a) Overlay of experimentally recorded time lapse for one beating cycle of an individual cilium (L = 480 μm). (b) Results from computer simulations obtained using the fluid-structure interaction model using COMSOL multiphysics. The actuation frequency is 1 Hz. Copyright S. Hanasoge, PJ. Hesketh, A. Alexeev, 2018.

small red arrows, is only significant near the cilium base. As the magnetic field rotates in the CCW direction, the local angle θ between the cilium and magnetic field direction increases, enhancing the magnitude of the magnetic moment. As a result of this CCW rotation, the cilium deforms and bends in the CCW direction.

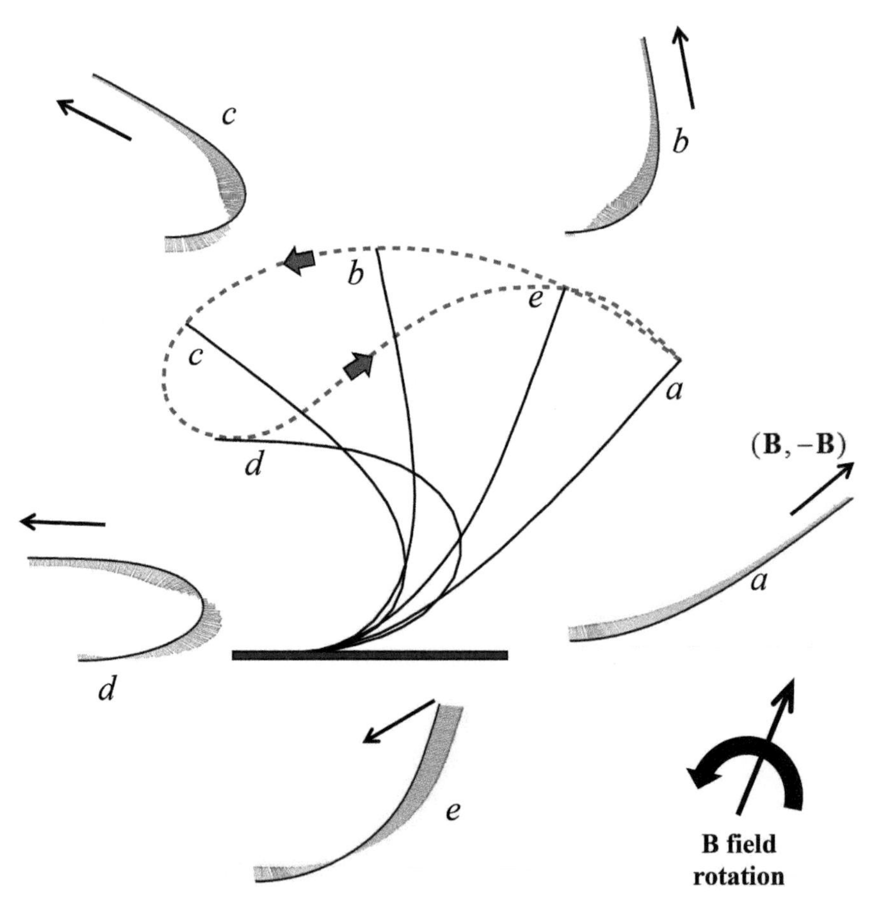

Figure 7.6 Cilium positions a–e at selected times throughout one oscillation cycle. Black arrows indicate the direction of the external magnetic field. The magnitude and direction of the applied magnetic moment along the cilium length are shown in red as a distribution plotted normal to the local cilium axis for the corresponding positions. Note that the magnetic moment varies continuously along the length. Copyright S. Hanasoge, PJ. Hesketh, A. Alexeev, 2018.

The cilium proceeds through positions b–c–d in Figure 7.2b following the rotation of the external magnetic field. The bending results in the local angle θ to exceed 90^0 near the cilium base. In position c, the base of the cilium makes an angle $\theta > 90^0$, whereas the cilium tip that more easily deflects to follow the magnetic field makes an angle $\theta < 90^0$. This results in a situation in which the direction of the local magnetic moment near the base is opposite to that at the cilium tip (Figure 7.6c). This change in the moment direction along cilium further increases the deformation.

The bending of cilium from *a* to *d* increases the elastic energy stored by cilium, which reaches a maximum at position *d*, where the cilium reaches the maximum bending. Further rotation of the magnetic field reduces the applied moment and the cilium rapidly recovers to the initial position *a* via position *e*, releasing the accumulated elastic energy. During the recovery stroke, the magnetization of the cilium flips the direction and so does the applied moment, which now acts to pull the cilium toward $-\mathbf{B}$ (see Figure 7.6e). Due to the fast cilium motion during the recovery stroke, the viscous force acting on the cilium is significantly higher than that during the forward stroke where the velocity is defined by the rate of magnetic field rotation. Furthermore, the magnetic moments flip direction during recovery and thus aid in the return of cilium to its initial position. As a result, the time scale for the recovery stroke is much shorter than that of the forward stroke, which in turn leads to an asymmetric beating. As magnetization flips during the recovery stroke, the cilium realigns with $-\mathbf{B}$ and repeats the motion. This leads to two beating cycles of the cilium for every magnet rotation.

7.4.2 Clockwise Rotation of Magnetic Field

The beating pattern of an elastic cilium driven by a magnetic field that rotates in the CW direction is shown in Figure 7.7. The cilium motion obtained in the experiments (Figure 7.7a) is well captured by the computer simulations (Figure 7.7b). Cilium beating differs significantly from the motion induced by the CCW rotation of the magnet (Figure 7.5). We define the forward stroke to begin at position *a* in Figure 7.7b, where the tip angle is close to perpendicular to the substrate. The cilium follows the direction of the magnetic field $\mathbf{B}$ as it rotates CW in the forward stroke. At the end of the forward stroke, the cilium touches the substrate and further bending is hindered by the substrate (position *b* in Figure 7.7b). The cilium remains in this position until the magnet rotates to a point, such that the magnetization within the cilium flips the direction. As soon as the magnetization direction flips, the magnetic moments act to deflect the cilium away from the substrate, performing a recovery stroke that aligns the cilium with $-\mathbf{B}$ in position *a*. During the recovery stroke, the cilium moves in the direction opposite to the rotation of the magnetic field.

Similar to CCW field rotation scenario, the cilium speed during the forward stroke is set by the rate of the magnetic field rotation. During the

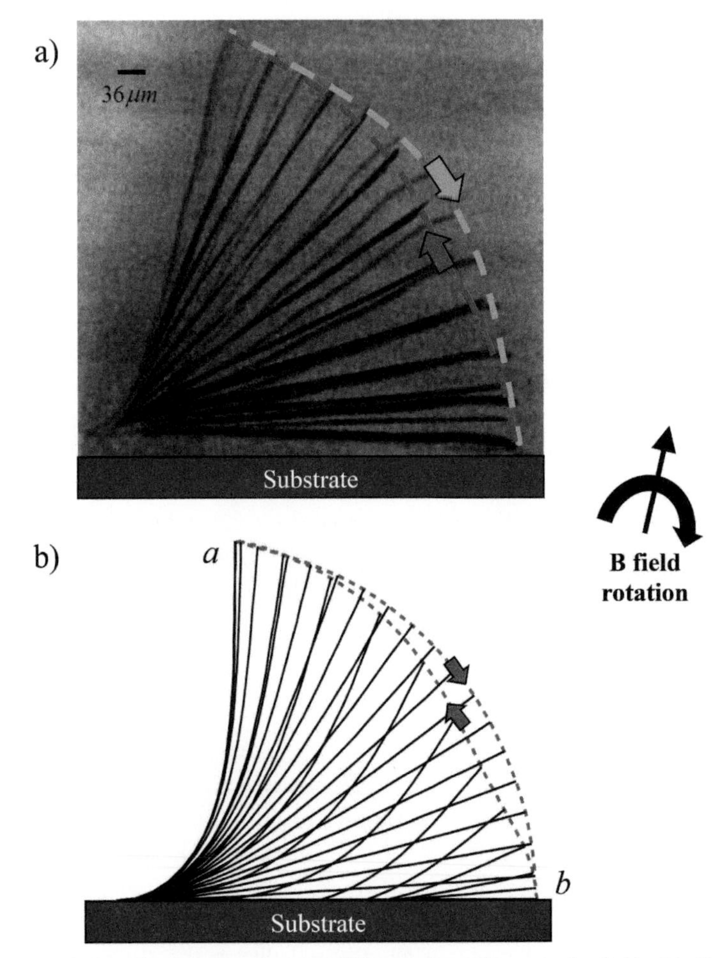

Figure 7.7 Cilium kinematics for the CW rotation of magnetic field. (a) Overlay of experimentally recorded time lapse. (b) Computer simulations obtained using COMSOL multiphysics. The cilium length is L = 480 μm, and the actuation frequency is 1 Hz. Copyright S. Hanasoge, PJ. Hesketh, A. Alexeev, 2018.

backward stroke, the cilium motion is defined by the balance between the magnetic force and the viscous force. Recall that for CCW rotation, the recovery stroke is the result of a balance between the elastic and viscous forces. The difference in the acting forces between the forward and recovery strokes gives rise to an asymmetry of the cilium motion in CW rotating magnetic field. Note that at the end of the forward stroke, the cilium pauses while the magnetic field continues to rotate CW until the magnetization in the cilium changes its direction. No such pause occurs for cilium driven by the CCW rotating magnetic field.

Both CW and CCW modes of cilium actuation result in two cycles of cilium beating for every rotation of the magnetic field as the cilium aligns with $\mathbf{B}$ and $-\mathbf{B}$. However, the kinematics is significantly different for these two modes. The CCW rotating magnetic field acts to increase the bending in the cilium during the forward stroke, whereas CW rotation reduces the bending in the forward stroke. These differences result in significantly different bending patterns for the CW and CCW directions of magnet rotation. Furthermore, CCW rotation leads to a noticeable larger spatial asymmetry of the beating pattern compared to the CW case.

Spatial asymmetry in a beating pattern is essential for creating a net fluid transport in low Reynolds number environments [15]. Indeed, a direct correlation between the area enclosed by the cilium tip trajectory and fluid transport has been suggested [17]. Thus, a greater fluid transport by cilia actuated by CCW field can be expected compared to CW actuation for similar operating conditions. Indeed, up to three times faster flows are reported for a ciliary array with CCW actuation compared to the same array actuated by a CW rotating magnetic field [19]. Furthermore, the direction of the fluid flow for the CW and CCW rotation of the field is found to be opposite [19]. A detailed discussion of kinematics of cilium for CCW magnet rotation and the dependence of such a motion on dimensionless parameters characterizing the system can be found elsewhere [15, 20].

7.5 Metachronal Motion

It has been observed that natural cilia often do not beat simultaneously. Metachronal motion is observed on a variety of spatial scales from cellular level [28–30] to larger scales in the gait of millipedes, spiders, and other invertebrates [31]. Researchers have identified specific advantages related to an increase in efficiency and performance for the use of metachronal motion by natural cilia [17, 29, 30, 32–35]. These advantages can be exploited in engineering systems by creating synthetic cilia capable of performing a similar metachronal motion [36].

Metachronal wave motion in a ciliary array emerges when there is a phase difference in beating strokes of neighboring cilia, such that neighboring cilia transition from the forward to the recovery stroke in a sequential manner. As discussed in the previous sections, magnetic cilia transition to the recovery stroke when the elastic force in the deformed cilia exceeds the magnetic force due to the rotating magnet. More flexible cilia deflect more easily by the magnetic field and, therefore, transition to the recovery at larger rotational angles of the magnetic field compared to stiffer cilia with the same magnetic properties. We can therefore use cilium stiffness to create a phase difference in the beating cycles.

The effective stiffness of cilia can be altered by changing cilia thickness, modulus, or length. In our system, we use the latter property to create arrays of magnetic cilia capable of metachronal motion [37]. We fabricate arrays of cilia with linearly increasing lengths and examine their behavior subject to a uniform rotating magnetic field. All the cilia in the arrays are 10 μm in width, and 60 nm in thickness. The length of the shortest cilium is 60 μm, and it linearly increases along the array up to 600 μm for the longest cilium.

Figure 7.8 presents snapshots of the fabricated arrays of magnetic cilia that are subjected to a uniform magnetic field. The figure shows the state of the array as the magnet rotates in the CCW direction. As the magnet rotates, the short cilia in the left side of the array completes the beating cycle first followed by longer cilia. The longest ones near the right side of the array complete the cycle last, after which the motion repeats. Thus, we find that a wave, with its front defined by the cilium that just completed the cycle, propagates along the cilium array. The rate of this metachronal wave is defined by the gradient of the cilium length along the array. The metachronal motion shown in this experiment is perpendicular to the direction of effective stroke (i.e., leoplectic), as the cilia of different lengths are arranged one next to the other. By changing the arrangement of cilia on the substrate, it is possible to create symplectic and antiplectic metachronal motion.

7.6 Microfluidic Applications of Artificial Cilia

Microscale fluid handling is essential for a variety of biomedical and other applications of the microfluidic technology. Researchers have recently reported such applications of microfluidics as diagnostics using liquid biopsy, real-time polymerase chain reaction, and single cell studies [38–42]. In this section, we demonstrate the use of artificial magnetic cilia for active fluid mixing and bacteria capture in microfluidic channels.

7.6.1 Microfluidic Mixing

Fluid mixing at the microscale is a challenging task because the flow is typically viscosity dominated and the mixing process is governed by a relatively

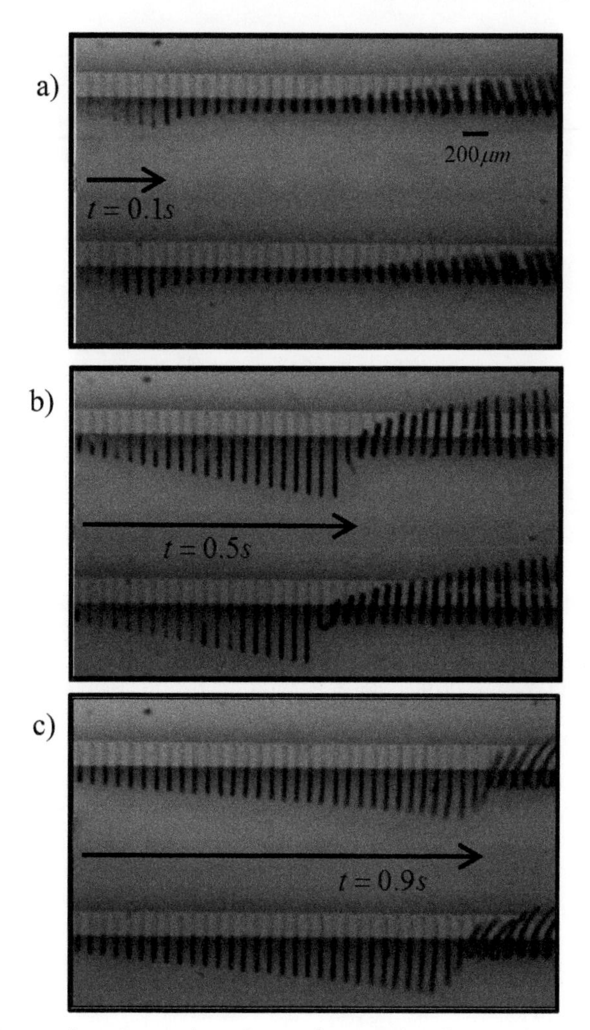

Figure 7.8 Propagation of metachronal wave in an array of magnetic cilia with different length. Arrow indicates the position of metachronal wave front that propagates from the left to the right. Short cilia on the left are actuated first, sequentially followed by the longer ones. The magnet is rotated in the CCW direction with frequency 1 Hz. Adapted from [37].

slow molecular diffusion. To achieve an efficient and fast mixing, fluid flows that cause repeated stretching and folding of the fluid have to be used [43, 44].

To examine the ability of magnetic cilia to effectively mix fluid streams, we use a microfluidic channel with an array of cilia actuated by a rotating

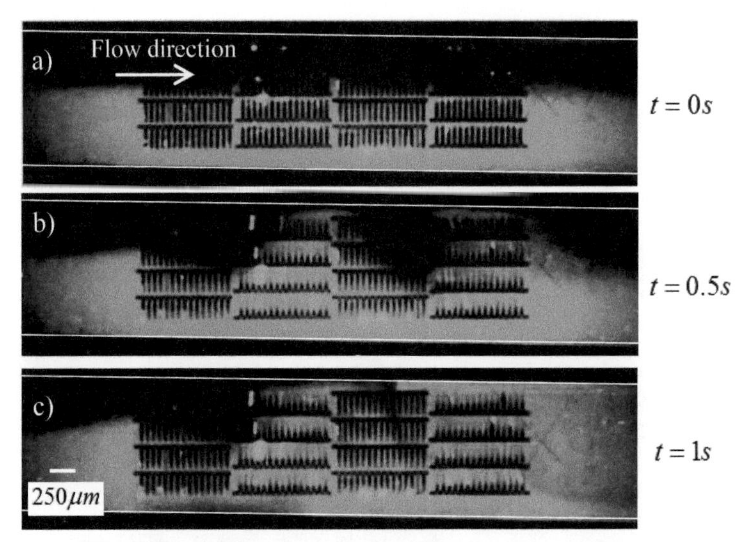

Figure 7.9 Mixing produced in a microchannel with two fluid streams. The inlet on the left shows two streams of fluid. The flowrate in the microchannel is 10 μl/min. (a) Cilia are not active, remaining flow streams unmixed. (b) Cilia are actuated at 50 Hz leading to the merging of the two fluid streams. (c) Steady-state flow in the channel with actuated cilia results in highly mixed streams of fluid downstream. Copyright S. Hanasoge, PJ. Hesketh, A. Alexeev, 2018.

permanent magnet. Two fluid streams, one of which is fluorescent, are introduced in a Y-shaped microchannel made by PDMS molding and bonding on a glass slide. The glass slide is decorated with ciliary array consisting of four columns with alternating pumping directions, as discussed after Figure 7.3.

Figure 7.9 shows experimental snapshots in which an array of cilia is used to mix two fluid streams pumped thorough the microchannel. Without any actuation of the ciliary array, two streams continue to flow nearly undisturbed along the array without intermixing, which indicates a minor effect of the diffusion on stream mixing (Figure 7.9a). As soon as the array is activated by the rotating permanent magnet, the cilia generate local flows normal to the direction of the fluid streams. Furthermore, the directions of the cilium induced flows alternate in the consecutive array columns resulting in flow circulations across the microchannel. These flow circulations have an immediate effect onto the two fluid steams by disrupting them and causing them to penetrate each other, as shown in Figure 7.9b. After a short transient, the flow within the microchannel reaches a steady state with a nearly complete mixing of the fluorescent and non-fluorescent streams in the microchannel downstream, as shown in Figure 7.9c. In our system, the mixing time is less than 1 s with mixing length much shorter than that required by diffusive mixing. Thus, the array of magnetic cilia demonstrates high potential for continuous mixing of fluids in microchannels.

7.6.2 Bacteria Capture

Isolating rare cells and pre-concentrating target species for analysis are important steps in the detection and enumeration of bacteria [45, 46] and cancer cells [47–49]. A typical microfluidic approach for bioparticle capture relies on target specific antibody coatings on passive surfaces in a microchannel that are used to arrest the target cells dispersed in a fluid stream [50–53]. The efficiency of capture in such systems depends on the probability of target cells coming in contact with the antibody-coated surface. The probability can be enhanced by vigorous fluid agitation near the coated surfaces.

We use antibody immobilized magnetic cilia to capture bacteria in a microfluidic channel flow. The cilia act to mix fluid and to create flow circulations that bring bacteria closer to the cilium surface. Furthermore, antibody-coated beating cilia sweep through a large area of fluid enhancing contact probability with target bacteria to facilitate their capture. Thus, the use of arrays of magnetic cilia can drastically increase the capture efficiency of the target compared to passive surfaces.

In our experiments, we capture *Salmonella enterica* using *anti-Salmonella*-coated cilia. The cells were labeled with Clontech Gfpuv green-fluorescent protein plasmid (Mountain View, CA, USA) using a calcium chloride heat shock transformation method [54]. To create the capturing cilia, we immobilize antibody-coated microbeads on the surface on the cilia. To this end, the cilia are first immersed in 1% APTES solution. The silane group binds to the thin oxide layer on the NiFe, and the amine group is protonated $-\mathrm{NH}_3^+$ creating an overall positive charge on the cilia. Next, carboxylate microbeads coated with *anti-Salmonella* are introduced. The beads with the carboxyl group $-COO^-$ have a slight negative charge and, therefore, bind to the surface of the cilia. This procedure results in cilia covered with antibody-coated microbeads that are electrostatically bound to the surface [55]. Similarly, the bottom glass wall of the microchannel is covered with antibody-coated microbeads. The latter is used to compare capturing efficiency between actively beating cilia and a passive static surface.

To evaluate capture of bacteria by magnetic cilia, we introduced *S. enterica* cells with a concentration of 10^5 cells/ml in a microchannel with an array

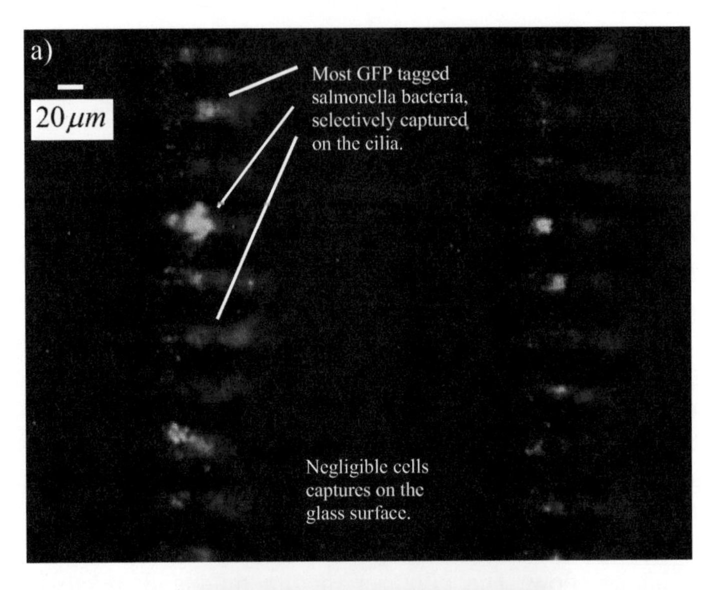

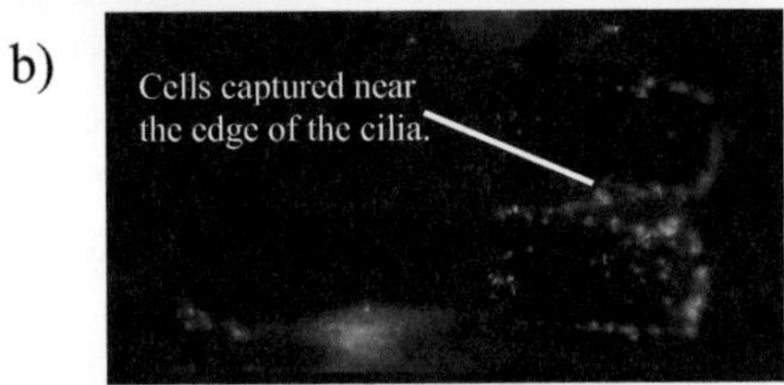

Figure 7.10 Bacteria capture with antibody immobilized cilia. (a) The GFP tagged fluorescent salmonella are selectively captured by the cilia. Very few cells attach to the bottom glass substrate. (b) Close up of the captured salmonella cells on the cilia surface. Notice that the cells stick mostly to the edges of the cilium. Copyright S. Hanasoge, Y. Ortega, M. Erikson, PJ. Hesketh, A. Alexeev, 2018.

of cilia attached to the bottom channel wall. We use a ciliary array which is identical to that shown in Figure 7.3. The cilia are actuated at 50 Hz for a period of 10 min during which 1 ml of the bacteria sample flows through across the array. Figure 7.10a shows images obtained using a fluorescent microscope that visualize the location of captured fluorescent bacteria on the cilia and channel substrate. The experiment reveals that the amount of *S. enterica* captured by the cilia is significantly larger than the amount of *S. enterica* captured on the bottom glass surface. Indeed, only a few cells can be seen sticking to the glass, whereas most of the cells are found on the cilia. Furthermore, the captured cells are localized on the edges of beating cilia where the highest local shear rates can be expected (Figure 7.10b). Thus, the experiment demonstrates the potential of using magnetic cilia to capture target species from a fluid for their effective isolation and pre-concentration.

7.7 Summary

We discussed the use of microscopic metallic thin films as artificial cilia with a biomimetic beating pattern. We showed that the cilia undergo an asymmetric beating leading to a net fluidic transport in a low Reynolds number inertia-less environment. We demonstrated that arrays of magnetic cilia can be designed to exhibit metachronal wave motion by altering the length of individual cilia. Finally, we discussed the applications of magnetic thin-film cilia for fluid mixing and bacteria capture.

Acknowledgments

We thank the USDA NIFA (grant #11317911) and the National Science Foundation (CBET-1510884) for financial support and the staff of Georgia Tech IEN for assistance with clean room fabrication. We thank Dr. Ynes Ortega and Dr. Marilyn Erickson from the University of Georgia for their help with the experiments with *S. enterica* bacteria.

References

[1] Kovarik ML, Gach PC, Ornoff DM, et al. Micro total analysis systems for cell biology and biochemical assays. *Anal. Chem.,* 84:516–540, 2012.
[2] Taylor G. Analysis of the swimming of microscopic organisms. In *Proceedings of the Royal Society of London A: Mathematical, Physical and Engineering Sciences,* 209:447–461, 1951.
[3] Purcell EM. Life at low Reynolds number. *Am. J. Phys.,* 45:3–11, 1977.
[4] Qiu T, Lee TC, Mark AG, et al. Swimming by reciprocal motion at low Reynolds number. *Nat. Commun.,* 5:5119, 2014.
[5] Turner L, Ryu WS, and Berg HC. Real-time imaging of fluorescent flagellar filaments. *J. Bacteriol.,* 182:2793–2801, 2000.
[6] Brokaw CJ. Non-sinusoidal bending waves of sperm flagella. *J. Exp. Biol.,* 43:155–169, 1965.

[7] Ballard M, Mills ZG, Beckworth S, et al. Enhancing nanoparticle deposition using actuated synthetic cilia. *Microfluid. Nanofluidics,* 17:317–324, 2014.

[8] Mills ZG, Aziz B, and Alexeev A. Beating synthetic cilia enhance heat transport in microfluidic channels. *Soft Matter,* 8:11508–11513, 2012.

[9] Masoud H, and Alexeev A. Harnessing synthetic cilia to regulate motion of microparticles. *Soft Matter,* 7:8702–8708, 2011.

[10] Branscomb J, and Alexeev A. Designing ciliated surfaces that regulate deposition of solid particles. *Soft Matter,* 6:4066–4069, 2010.

[11] Ghosh R, Buxton GA, Usta OB, et al. Designing oscillating cilia that capture or release microscopic particles. *Langmuir,* 26:2963–2968, 2010.

[12] Toonder JM, den J, and Onck PR. Microfluidic manipulation with artificial/bioinspired cilia. *Trends Biotechnol.,* 31:85–91, 2013.

[13] Gorissen B, de Volder M, and Reynaerts, D. Pneumatically-actuated artificial cilia array for biomimetic fluid propulsion. *Lab. Chip,* 15:4348–4355, 2015.

[14] Masuda T, Akimoto AM, Nagase K, et al. Artificial cilia as autonomous nanoactuators: Design of a gradient self-oscillating polymer brush with controlled unidirectional motion. *Sci. Adv.,* 2:e1600902, 2016.

[15] Khaderi SN, Baltussen MGHM, Anderson PD, et al. Breaking of symmetry in microfluidic propulsion driven by artificial cilia. *Phys. Rev. E,* 82:027302, 2010.

[16] Wang Y, den Toonder J, Cardinaels R, et al. A continuous roll-pulling approach for the fabrication of magnetic artificial cilia with microfluidic pumping capability. *Lab. Chip,* 16:2277–2286, 2016.

[17] Khaderi SN, Craus CB, Hussong J, et al. Magnetically-actuated artificial cilia for microfluidic propulsion. *Lab. Chip,* 11:2002–2010, 2011.

[18] Roper M, Dreyfus R, Baudry J, et al. On the dynamics of magnetically driven elastic filaments. *J. Fluid Mech.,* 554:167, 2006.

[19] Hanasoge S, Hesketh PJ, and Alexeev A. Microfluidic pumping using artificial magnetic cilia. *Microsyst. Nanoeng.*, 4:11, 2018. doi:10.1038/s41378-018-0010-9

[20] Hanasoge S, Ballard MJ, Hesketh P, et al. Asymmetric motion of magnetically actuated artificial cilia. *Lab. Chip,* 17:3138–3145, 2017.

[21] Vilfan M, Potoènik A, Kavèiè B, et al. Self-assembled artificial cilia. *Proc. Natl. Acad. Sci.,* 107:1844–1847, 2010.

[22] Belardi J, Schorr N, Prucker O, et al. Artificial cilia: generation of magnetic actuators in microfluidic systems. *Adv. Funct. Mater.,* 21:3314–3320, 2011.
[23] Sun L, and Zheng Y. Bio-inspired artificial cilia with magnetic dynamic properties. *Front. Mater. Sci.,* 9:178–184, 2015.
[24] Fahrni F, Prins MWJ, and van Ijzendoorn LJ. Micro-fluidic actuation using magnetic artificial cilia. *Lab. Chip,* 9:3413, 2009.
[25] den Toonder J, Bos F, Broer D, et al. Artificial cilia for active microfluidic mixing. *Lab. Chip,* 8:533, 2008.
[26] Hanasoge S, Owen D, Ballard M, et al. Active fluid mixing with magnetic microactuators for capture of Salmonella. In *Sensing for Agriculture and Food Quality and Safety VIII*, eds. Kim, M. S., Chao, K., and Chin, B. A., 986405, 2016.
[27] Babataheri A, Roper M, Fermigier M, et al. Tethered fleximags as artificial cilia. *J. Fluid Mech.,* 678:5–13, 2011.
[28] Gueron S, Levit-Gurevich K, Liron N, et al. Cilia internal mechanism and metachronal coordination as the result of hydrodynamical coupling. *Proc. Natl. Acad. Sci.,* 94:6001–6006, 1997.
[29] Elgeti J, and Gompper G. Emergence of metachronal waves in cilia arrays. *Proc. Natl. Acad. Sci.,* 110:4470–4475, 2013.
[30] Guirao B, and Joanny J-F. Spontaneous creation of macroscopic flow and metachronal waves in an array of cilia. *Biophys. J.,* 92:1900–1917, 2007.
[31] Wilkinson M. *The Story of Life in Ten Movements*. Basic Books, 2016.
[32] Osterman N, and Vilfan A. Finding the ciliary beating pattern with optimal efficiency. *Proc. Natl. Acad. Sci.,* 108:15727–15732, 2011.
[33] Khaderi SN, den Toonder JMJ, and Onck PR. Fluid flow due to collective non-reciprocal motion of symmetrically-beating artificial cilia. *Biomicrofluidics,* 6:014106, 2012.
[34] Mayne R, Whiting JGH, Wheway G, et al. Particle sorting by Paramecium cilia arrays. *Biosystems,* 156–157:46–52, 2017.
[35] Georgilas I, Adamatzky A, Barr D, et al. Metachronal waves in cellular automata: cilia-like manipulation in actuator arrays. In *Nature Inspired Cooperative Strategies for Optimization (NICSO 2013)* (Springer, Cham), 261–271, 2014. doi:10.1007/978-3-319-01692-4_20
[36] Tsumori F, Marume R, Saijou A, et al. Metachronal wave of artificial cilia array actuated by applied magnetic field. *Jpn. J. Appl. Phys.,* 55:06GP19, 2016.

[37] Hanasoge S, Hesketh PJ, and Alexeev A. Metachronal motion of artificial magnetic cilia. *Soft Matter,* 14:3689–3693, 2018. doi:10.1039/C8SM00549D

[38] Brock G, Castellanos-Rizaldos E, Hu L, et al. Liquid biopsy for cancer screening, patient stratification and monitoring. *Transl. Cancer Res.,* 4:280–290, 2015.

[39] Neoh KH, Hassan AA, Chen A, et al. Rethinking liquid biopsy: Microfluidic assays for mobile tumor cells in human body fluids. *Biomaterials,* 150:112–124, 2018.

[40] Zhang Y, and Jiang H-R. A review on continuous-flow microfluidic PCR in droplets: Advances, challenges and future. *Anal. Chim. Acta,* 914:7–16, 2016.

[41] Ahrberg D, Manz C, and Chung G. Polymerase chain reaction in microfluidic devices. *Lab. Chip,* 16:3866–3884, 2016.

[42] Reece A, Xia B, Jiang Z, et al. Microfluidic techniques for high throughput single cell analysis. *Curr. Opin. Biotechnol.,* 40:90–96, 2016.

[43] Lee C-Y, Chang C-L, Wang Y-N, et al. Microfluidic Mixing: A Review. *Int. J. Mol. Sci.,* 12:3263–3287, 2011.

[44] Owen D, Ballard M, Alexeev A, et al. Rapid microfluidic mixing via rotating magnetic microbeads. *Sens. Actuators Phys.,* 251:84–91, 2016.

[45] Liu C, Lagae L, and Borghs G. Manipulation of magnetic particles on chip by magnetophoretic actuation and dielectrophoretic levitation. *Appl. Phys. Lett.,* 90:184109, 2007.

[46] Wu Z, Willing B, Bjerketorp J, et al. Soft inertial microfluidics for high throughput separation of bacteria from human blood cells. *Lab. Chip,* 9:1193–1199, 2009.

[47] Gascoyne PRC, Noshari J, Anderson TJ, et al. Isolation of rare cells from cell mixtures by dielectrophoresis. *Electrophoresis,* 30:1388–1398, 2009.

[48] Nagrath S. Sequist LV, Maheswaran S, et al. Isolation of rare circulating tumour cells in cancer patients by microchip technology. *Nature,* 450:1235–1239, 2007.

[49] Pratt ED, Huang C, Hawkins BG, et al. Rare cell capture in microfluidic devices. *Chem. Eng. Sci.,* 66:1508–1522, 2011.

[50] Gleghorn JP, Pratt ED, Denning D, et al. Capture of circulating tumor cells from whole blood of prostate cancer patients using geometrically enhanced differential immunocapture (GEDI) and a prostate-specific antibody. *Lab. Chip,* 10:27–29, 2010.

[51] Ji HM, Samper V, Chen Y, et al. Silicon-based microfilters for whole blood cell separation. *Biomed. Microdevices,* 10:251–257, 2008.
[52] Wang Z, Chin SY, Chin CD, et al. Microfluidic CD4+ T-cell counting device using chemiluminescence-based detection. *Anal. Chem.,* 82:36–40, 2010.
[53] Murthy SK, Sin A, Tompkins RG, et al. Effect of flow and surface conditions on human lymphocyte isolation using microfluidic chambers. *Langmuir ACS J. Surf. Colloids,* 20:11649–11655, 2004.
[54] Ma L, Zhang G, and Doyle MP. Green fluorescent protein labeling of *Listeria*, *Salmonella*, and *Escherichia coli* O157:H7 for safety-related studies. *PLoS ONE,* 6:e18083, 2011.
[55] Sivagnanam V, Sayah A, Vandevyver C, et al. Micropatterning of protein-functionalized magnetic beads on glass using electrostatic self-assembly. *Sens. Actuators B Chem.,* 132:361–367, 2008.

8

Artificial Cilia: Fabrication, Actuation, and Flow Generation

Jaap M. J. den Toonder[1], Ye Wang[1], Shuaizhong Zhang[1] and Patrick Onck[2]

[1]Department of Mechanical Engineering and Institute of Complex Molecular Systems, Eindhoven University of Technology, Eindhoven, 5600 MB, The Netherlands
[2]Zernike Institute for Advanced Materials, University of Groningen, Groningen, 9747 AG, The Netherlands
E-mail: j.m.j.d.toonder@tue.nl; Y.Wang2@tue.nl; s.zhang1@tue.nl; p.r.onck@rug.nl

Control of fluid flow at small length scales (typically <1 mm), i.e., "microfluidics," is important for many applications. Examples are devices for healthcare diagnostics, in which complex tasks of (bio-)fluid manipulation and detection need to be performed, and organ-on-a-chip devices in which suitable micro-environments for multicellular tissue structures need to be created. Recently, researchers have started developing artificial cilia for microfluidic pumping and mixing. In this chapter, we describe the fabrication methods we have developed for realizing artificial cilia, discuss their actuated motion, and show the fluid flow they generate when they are integrated in microfluidic devices. We start with electrostatically actuated cilia, but the main focus is on artificial cilia that are actuated magnetically.

8.1 Introduction

Although microfluidics-based applications are being commercialized especially in medical diagnosis, the field of microfluidics is still much in development with many different technological approaches, materials, and principles being explored simultaneously. In the last two decades, much effort has been put into developing microfluidic components such as micropumps and micromixers. Most of these approaches need either large peripheral equipment, such as pneumatic control systems or syringe pumps, or expensive or cumbersome to integrate, such as electrode patterns for electro-hydrodynamic pumping.

Inspired by nature, artificial cilia have recently been developed to create pumping and/or mixing in microfluidic devices [1, 2]. A major potential of artificial cilia is that, in contrast to conventional micro-pumps and mixers, they can be integrated in a straightforward manner, and do not require large peripheral equipment to drive them. The first publications of artificial cilia appeared about a decade ago and since then, microscopic actuators resembling motile cilia, actuated to move fluid under the influence of a number of different stimuli such as electrostatic [3], magnetic [4–16], pneumatic [17], resonance-actuated [18, 19], light [20], and pH [21], have been developed by a number of groups. These artificial cilia cover parts of the inner walls of the microfluidic channels, sometimes arranged in specific patterns, and have been shown to be capable of generating flow and mixing in microfluidic environments. Among these, magnetic artificial cilia (MAC) are the most promising, because (1) they can be externally actuated within microfluidic channels by permanent magnets or electromagnets without the need for physical connections to peripheral equipment, (2) they have an instantaneous response to the external stimulus, and (3) magnetic actuation is compatible with biological fluids. To realize these artificial cilia, a wide variety of fabrication approaches have been used, ranging from thin-film techniques combined with lithography [4, 7, 9], self-assembly methods using magnetic particles aligning in magnetic fields [8, 10, 12], using polycarbonate track-etched porous membranes as sacrificial templates [5, 6], and roll-to-roll micropulling [13], to micro-molding [17, 22–24].

One of the first types of artificial cilia was our electrostatically actuated artificial cilia presented in [3] (see Figure 8.1). These were manufactured using microelectromechanical systems (MEMS) technology. Figure 8.1 shows the working principle: a thin bilayer of polymer (polyimide) and metal (chromium) is partially released from the substrate during the cilia fabrication process, and by controlling the stresses in the bilayer structure, a curved microflap is obtained. The right image in Figure 8.1 is an electron micrograph of the electrostatic artificial cilia, which have a length of 100 μm, a width of 20 μm, and a thickness of 1 μm. By applying a voltage difference between the chromium and an electrode embedded in the surface (indium tin oxide), an electrostatic force pulls the cilia down, and they roll out over the surface.

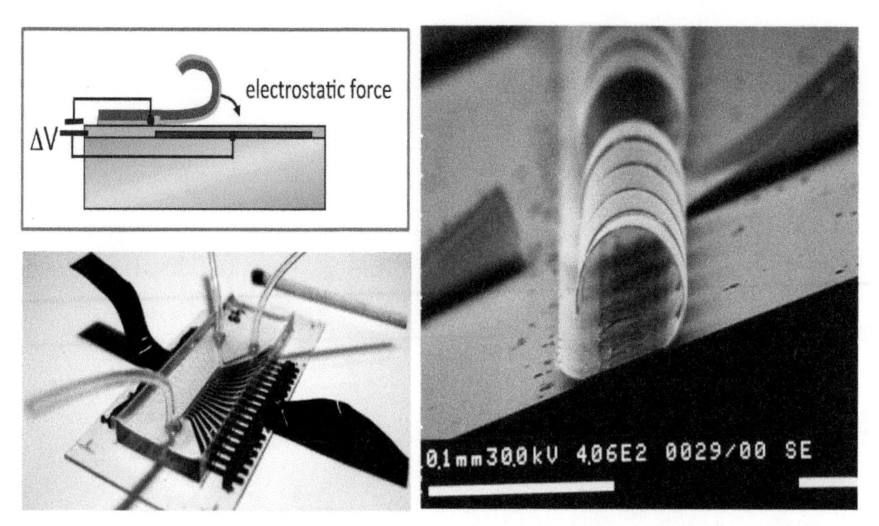

Figure 8.1 Electrostatic artificial cilia, manufactured using MEMS technology. The top left shows the working principle: a thin bilayer of polymer (polyimide) and metal (chromium) is partially released from the substrate, forming a curved flap. By applying a voltage difference between the chromium and an electrode embedded in the surface (indium tin oxide), the emerging electrostatic force pulls the cilium down, and it rolls out over the surface. If the voltage is released, the cilium springs back to its bent state due to elastic recovery. The bottom left shows a microfluidic device, in which the electrostatic cilia are integrated – on the microchannel floor present in the device. The right image is an electron micrograph of the electrostatic artificial cilia, which have a length of 100 μm, a width of 20 μm, and a thickness of 1 μm. Reproduced with permission from [3].

If the voltage is released, the cilia spring back to their curved state due to elastic recovery.

These electrostatic cilia were integrated into a microfluidic device (shown in the bottom left in Figure 8.1), on the microchannel bottom surface of the device, in specific patterns. Figure 8.2 shows two of these, designed to function as a micromixer. The channel runs from left to right over the patterns. The thick red arrows indicate the fluid flow direction induced through the channels by an external pump. In each of the 1 $\times$ 1 mm square segments, five rows of 20 artificial electrostatic cilia are present. The thin red arrows indicate the rolling direction of the cilia. The top and the bottom images in Figure 8.2 are two different designs. As the fluid is pumped through the microchannel (from left to right) the moving artificial cilia generate a secondary flow (over the channel cross section), which changes in nature each time when the fluid

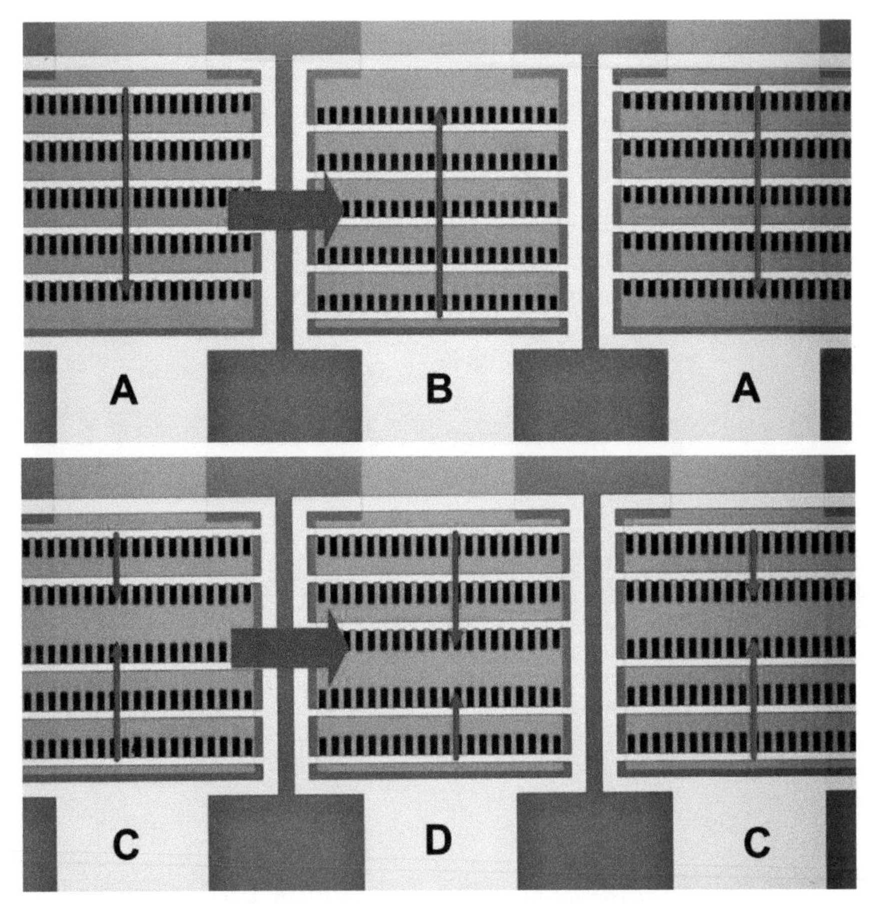

Figure 8.2 Top-view images of the microchannel floor of the device of Figure 8.1. The channel runs from left to right. The thick red arrows indicate the fluid flow direction induced by an external pump. In each of the squares shown, five rows of 20 artificial electrostatic cilia are present. The thin red arrows indicate the rolling direction of the cilia. The top and bottom are two different designs. Reproduced with permission from [3].

moves from one segment to the other. By design, the secondary patterns are expected to create chaotic advection, which leads to micromixing.

Indeed, Figure 8.3 shows that efficient mixing occurs. The figure includes snapshots from a mixing experiment using dyed silicone oils (viscosity 0.93 mPa.s) and mixing configuration design A–B–A from Figure 8.2. The externally driven mean velocity (from left to right) is 2 mm s^{-1}. At time $t = 0$ s, the artificial cilia are switched on with a frequency of 50 Hz at

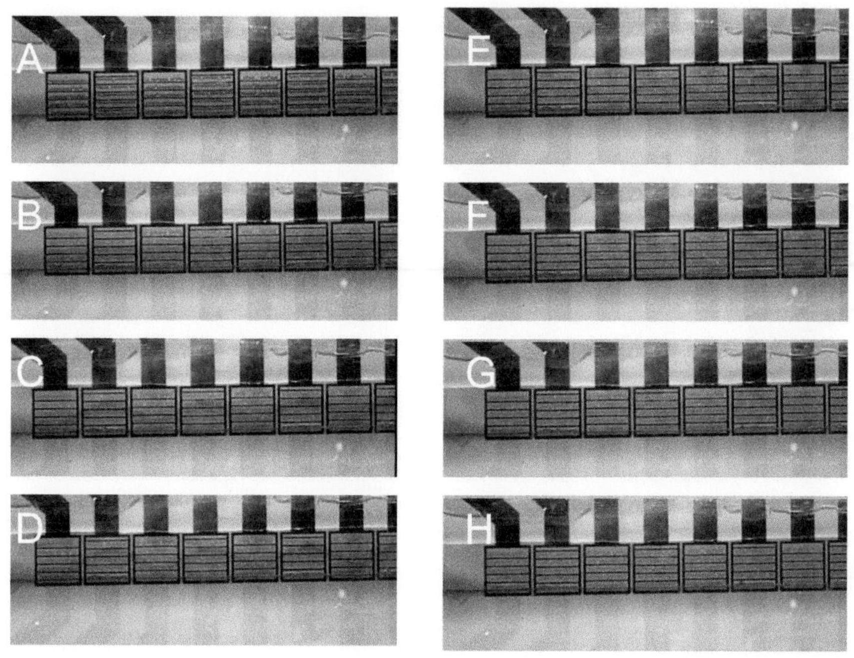

Figure 8.3 Snapshots from a mixing experiment using dyed silicone oils (viscosity 0.93 mPa.s) and mixing configuration design A–B–A from Figure 8.2. The externally driven main mean velocity (from left to right) is 2 mm s^{-1}. At time $t = 0$ s, the artificial cilia are switched on with a frequency of 50 Hz at an AC voltage of 100 V/1 kHz. (a) Not actuated. (b) $t = 0.00$ s. (c) $t = 0.04$ s. (d) $t = 0.08$ s. (e) $t = 0.12$ s. (f) $t = 0.16$ s. (g) $t = 0.44$ s. (h) $t = 1.44$ s. A meandering flow pattern almost immediately occurs, and within 1.5 s, the fluid is homogeneously mixed. This time corresponds to less than 1.5 cycles traveling distance in the main flow direction. Reproduced with permission from [3].

an AC voltage of 100 V/1 kHz. A meandering flow pattern almost immediately occurs, and within 1.5 s, the fluid is homogeneously mixed. This time corresponds to less than 1.5 cycles traveling distance in the main flow direction.

To understand the effects observed in these experiments, we carried out flow visualization measurements using optical coherence tomography (OCT) in combination with performing numerical simulations [25]. The main results are shown in Figure 8.4. Indeed, the cross-sectional flow exhibits the typical vortical motion that leads to mixing, as found from the OCT experiments. This behavior could be reproduced only in the simulations when assuming a Reynolds number greater than 1 – which means that we had to assume

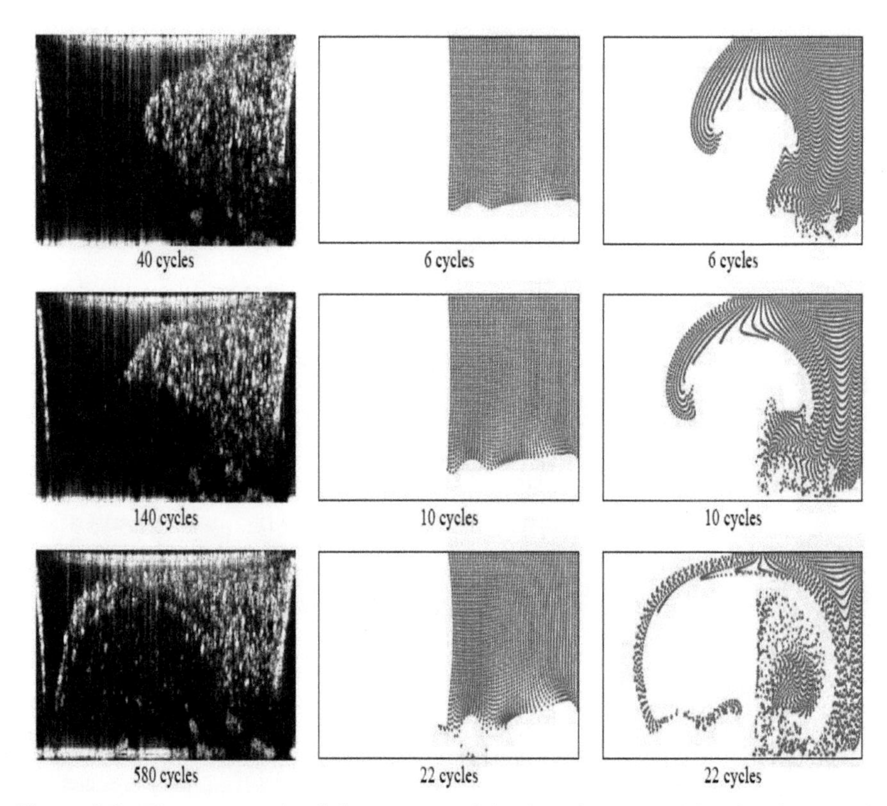

Figure 8.4 The cross-sectional flow generated in the microchannel by the electrostatic artificial cilia was characterized experimentally using OCT. Next to this, the flows were numerically simulated and observed by calculated particle positions over time. Left: images from OCT experiments – the cilia are on the bottom, attached to the floor of the channel; the cilia configuration is that of the CDC pattern in Figure 8.2. Middle: simulated particle distributions, with the assumption that the Reynolds number Re = 0. Right: simulated particle distributions for Re = 10. Reproduced with permission from [25].

that inertial effects must play a role. This is in fact a consequence of the very fast and symmetrical motion of the electrostatic artificial cilia during actuation. This type of behavior is in contrast with that of biological cilia, where scales are small and timescales relatively slow, and therefore inertia is always negligible. This is why biological cilia must move asymmetrically to generate a net flow.

Although the electrostatic artificial cilia can generate substantial flow speeds and work quite effectively as mixers, they have disadvantages. One of them is the fact that an electric field is by definition present within the

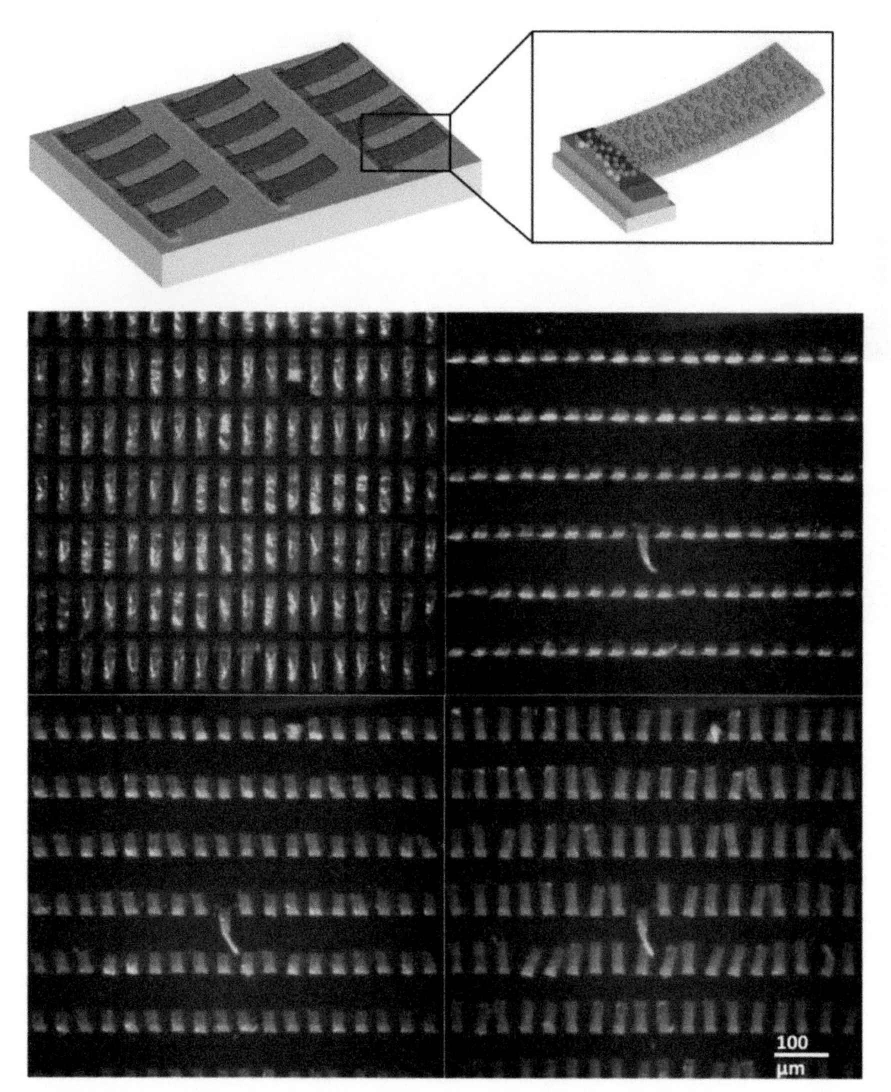

Figure 8.5 MAC. These were made with a lithography process, and consist of polymer flaps (poly(*n*-butylacrylate)) with embedded magnetic nanoparticles, and partially released from the surface [26, 27]. The length, width, and thickness of these MAC are 70, 20, and 0.4 μm, respectively. The cilia can be actuated using a rotating magnet. The optical micrographs show the cilia at various stages of actuation. Reproduced with permission from [26] and [27].

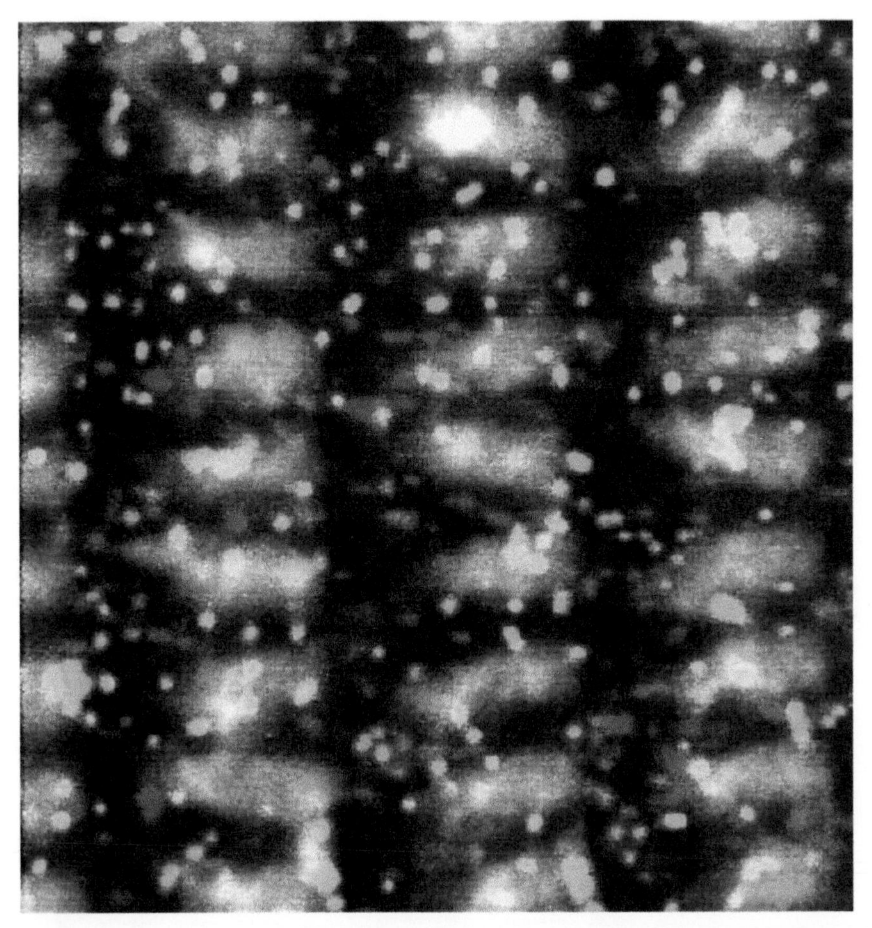

Figure 8.6 The flow generation capability of the MAC of Figure 8.5 was characterized using particle tracking. This figure shows the superposition of snapshots from two experiments (represented by differently colored particles); the flap-like cilia can be seen in the background as light-gray features. Flow velocities of more than 100 μm/s could be measured. Used with permission from J. Rühe. Copyright J. Rühe 2011.

microchannel, and the required voltages for the actuation of the electrostatic artificial cilia are relatively high. This leads to electrochemical effects and thus gas formation, particularly in fluids containing charge carriers (e.g., ions), which in fact prevents these artificial cilia from being used in aqueous liquids, and therefore they cannot be used in biological fluids like blood. Therefore, we turned next to MAC, which do not have this disadvantage.

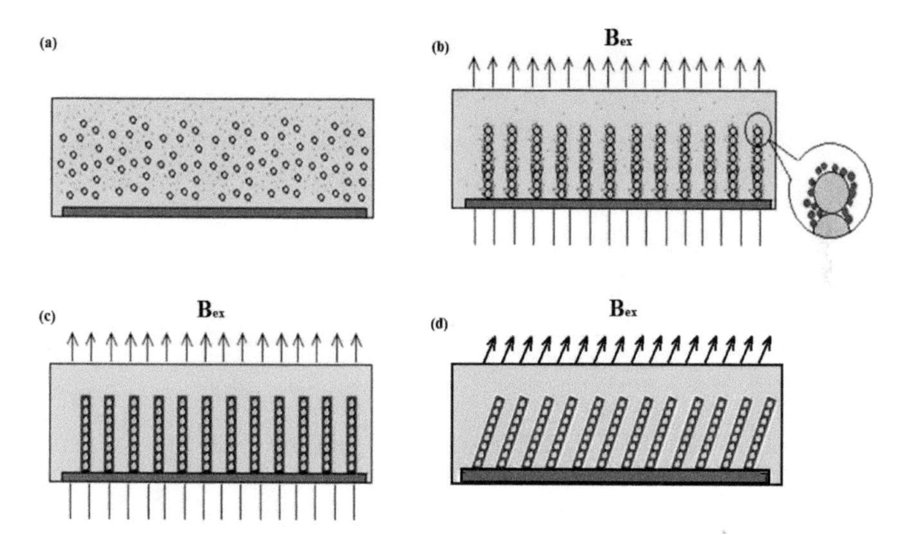

Figure 8.7 Fabrication scheme of MAC based on magnetic self-assembly. First, (a) a mixture of 2.7 μm spherical MBs and PBA latex nanoparticles is placed in a fluid cell. The black circles are MBs and red dots are PBA latex nanoparticles. Then (b) MBs are linked into chains and attached to the surface in the presence of an external magnetic field. Meanwhile (b) and (c), latex particles attach to the beads by electrostatic attraction and form a continuous layer around the chains after heating. Finally (d), the enclosing PBA layer is formed and the artificial cilia can be actuated by an external field. Reproduced with permission from [10].

Figure 8.5 shows MAC developed within a collaborative European project, at the lab of Prof. Rühe at IMTEK in Freiburg [26]. These were made with a lithography process, and consist of polymer flaps (poly(*n*-butylacrylate)) with embedded magnetic nanoparticles; during the process, they are partially released from the surface. The length, width, and thickness of these MAC are 70, 20, and 0.4 μm, respectively. The cilia can be actuated using a rotating magnet underneath the microfluidic channel in which the artificial cilia are integrated. The motion these artificial cilia make is similar to that of the electrostatic cilia discussed above, but the speed of their motion is lower and the stroke is non-symmetric. The optical micrographs in Figure 8.5 show the cilia at various stages of actuation. The motion of these artificial cilia can induce substantial fluid flow, as illustrated in Figure 8.6, and quantified in [8] and [27]: flow velocities of water of more than 100 μm/s could be generated.

A disadvantage of the MAC shown in Figures 8.5 and 8.6 is their intricate fabrication process, based on lithography and involving multiple steps. Our

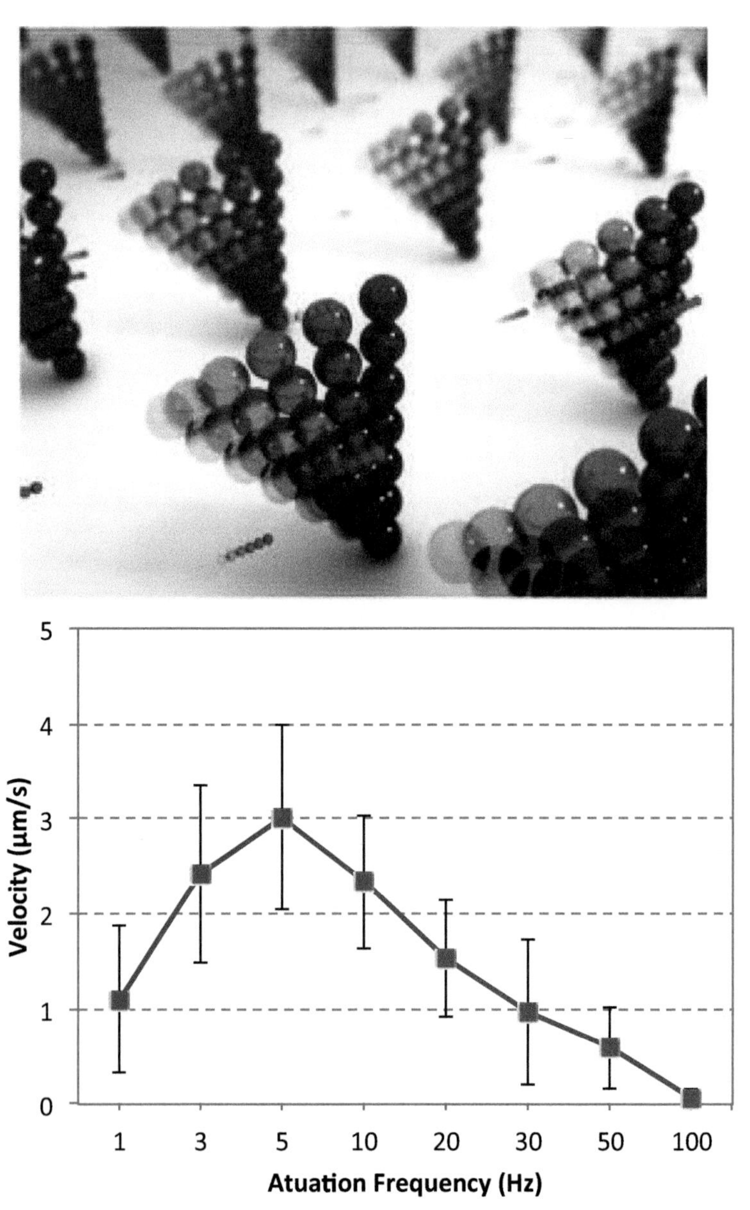

Figure 8.8 Top: Artist's impression of the motion of the self-assembled MAC, exhibiting a tilted conical motion in a rotating magnetic field. Bottom: The measured flow velocity generated by these magnetic cilia as a function of rotation frequency. Reproduced with permission from [10].

next goal was therefore to fabricate MAC using simpler, "out-of-cleanroom" methods.

Figure 8.7 shows the fabrication scheme of MAC based on magnetic self-assembly [10]. First, in Figure 8.7a, a mixture of 2.7 μm spherical magnetic beads (MBs) and poly(butyl acrylate) (PBA) latex nanoparticles is placed in a fluid cell. The black circles are MBs and red dots are PBA latex nanoparticles. Then in Figure 8.7b, MBs are linked into chains and attached to the surface in the presence of an external magnetic field. Meanwhile, in Figures 8.7b and c, latex particles attach to the beads by electrostatic attraction and form a continuous and permanent layer around the chains after heating. Finally in Figure 8.7d, the enclosing PBA layer is formed and the artificial cilia, formed by linked magnetic microparticles, can now be actuated by an external field. The field, generated by an electromagnetic octopole, resulted in the self-assembled MAC exhibiting a tilted conical motion as sketched in Figure 8.8. Such a motion is asymmetric with respect to the surface on which the cilia are placed, and therefore is expected to generate a net flow, even at low Re conditions in which inertia is absent. The bottom graph in Figure 8.8 depicts the measured flow velocity generated by the magnetic cilia as a function of rotation frequency. The velocities are relatively small, because of the small cilia size (i.e., a length of about 20–30 μm). Remarkably, the velocity has a maximum at a certain rotation frequency. This is due to the balance between magnetic and viscous forces acting on the cilia. The magnetic field creates a torque that makes the cilia rotate. At the same time, the motion of the cilia results in a viscous drag that acts to diminish the cilia motion, and this effect becomes stronger when the cilia rotate faster. Hence, at a critical rotation speed, the counteracting effect of the viscous drag dominates over the magnetic torque, resulting in the artificial cilia moving at a decreased amplitude (albeit faster), hence reducing the flow induced.

In [11], we introduced the "magnetic fiber drawing" technique to create MAC that are larger than the self-assembled cilia, and hence are expected to generate higher flow speeds. This technique starts with making a precursor film, consisting of uncured polymer (polydimethylsiloxane, PDMS) with dispersed magnetic microparticles. Then, a strong magnet is positioned at a small distance from this film, and due to the high magnetic field gradient generated by the magnet, the dispersed particles start lining up, and are drawn toward the magnet, dragging with them the uncured PDMS, to form thin fibers. The resulting artificial cilia are shown in Figure 8.9. By controlling the process parameters, the thickness, length, and cilia density could be controlled, as shown in the figure. After integration of these artificial cilia

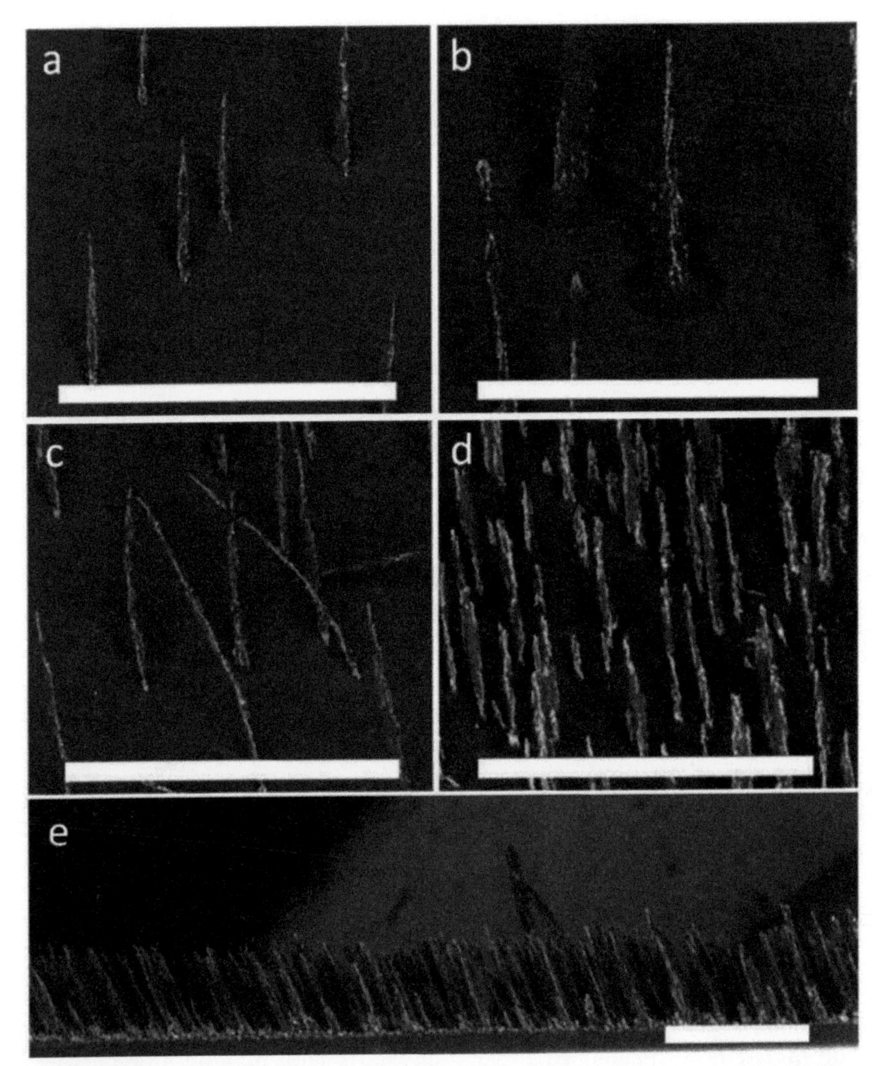

Figure 8.9 Controlling artificial cilia geometry and density using the "magnetic fiber drawing" technique introduced in [11]. The cilia shown in (a) were made with a magnetic particle concentration of 10 wt%, a precursor layer thickness of 60 μm, and a fiber drawing time of 10 s; increasing magnetic particle concentration to 20 wt% in the precursor resulted in thicker cilia (b); increasing thickness of the precursor layer to 120 μm resulted in an increase in length (c); and increasing the drawing time to 10 min resulted in a higher overall cilia density (d). (e) Side view of artificial cilia showing that a collection of artificial cilia with well-controlled maximum lengths can be produced. The scale bars in (a–d) are 500 μm. Reproduced with permission from [11].

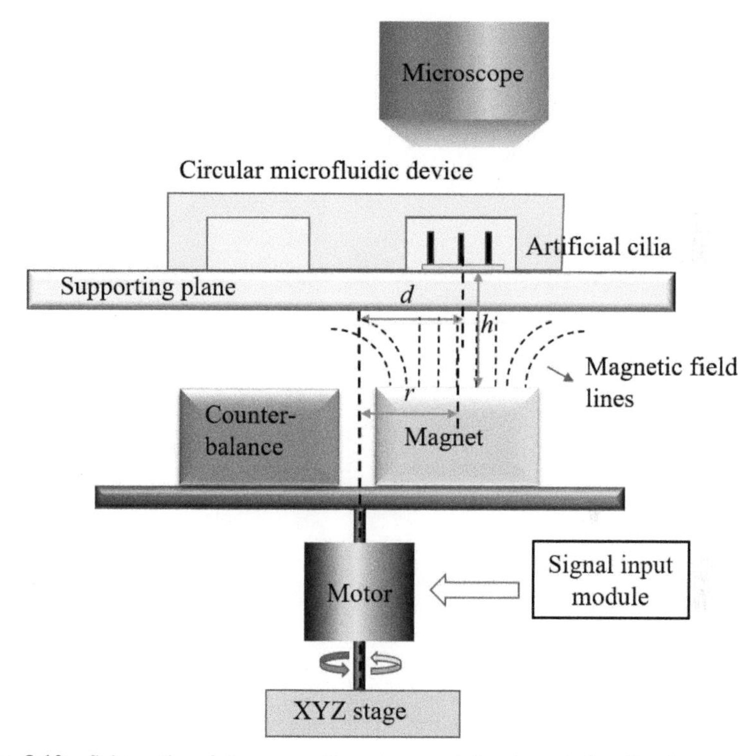

Figure 8.10 Schematic of the magnetic setup used to characterize the flow generation properties of MAC. The cilia are integrated on the floor of a microfluidic device. A permanent magnet is rotated underneath the device, with the rotation axis offset from the cilia location. This generates a time-dependent magnetic field that causes the cilia to perform rotation along a tilted cone. The cilia motion and the fluid flow can be observed from above using a microscope. Reproduced with permission from [13].

in a microfluidic chip, we used the setup shown in Figure 8.10 to actuate the cilia: a permanent magnet is rotated underneath the device, with the rotation axis offset from the cilia location. This generates a time-dependent magnetic field that causes the cilia to perform rotation along a tilted cone (which is expected to generate a net flow, like we discussed earlier). The cilia motion and the fluid flow can be observed from above using a microscope. Figure 8.11 shows how these artificial cilia move. Top-view images of the motion of the MAC are shown in water (Figure 8.11a) and glycerol (Figure 8.11b). Each image is a composition of 50 frames during one actuation cycle. A clear trend of diminishing in the size of the cone traced by artificial cilia during actuation at higher frequencies can be observed. This is due to the effect of

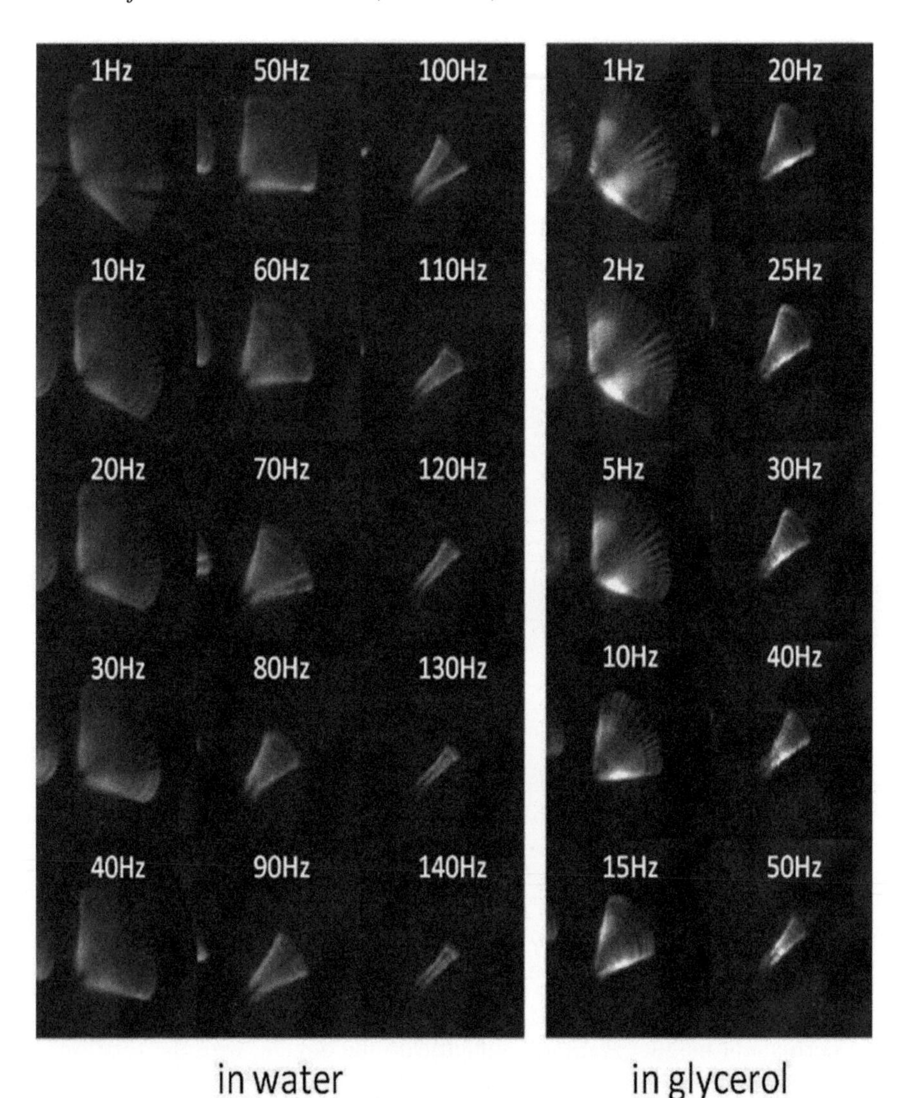

Figure 8.11 Top-view images of the motion of the MAC made using the magnetic fiber drawing process (Figure 8.9) captured with a high-speed camera during actuation using the setup of Figure 8.10, in (a) water and (b) glycerol. Each image is a composition of 50 frames during one actuation cycle. A clear trend of diminishing in the size of the cone traced by artificial cilia during actuation can be observed. Reproduced with permission from [11].

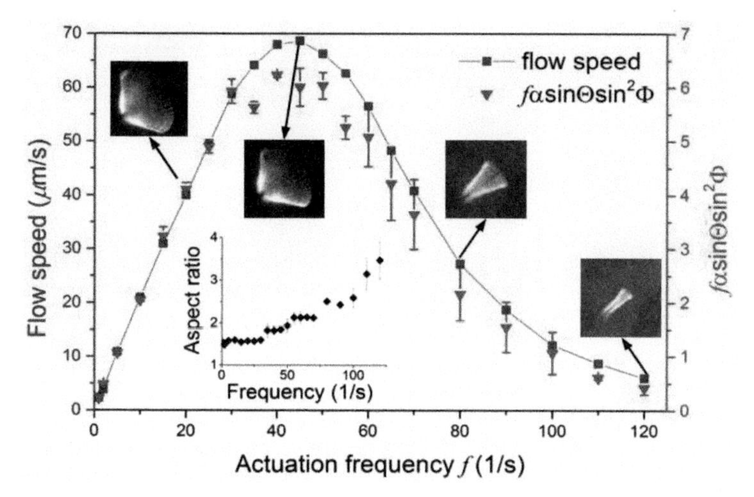

Figure 8.12 Flow characterization (water) and the dynamics of MAC made with magnetic fiber drawing. Measured flow speed is shown along with the theoretical prediction $f\alpha(\sin\theta\sin^2\Phi)$ at different actuation frequencies f fitted by scaling, in which θ is the cone tilt angle and Φ is the cone opening angle measured on the short axis of the ellipsoidal shaped cone. Change of the aspect ratio of the ellipsoidal cone at different frequencies is shown in the embedded graph. Snapshots at various frequencies show the change in the trace of cilia during one actuation cycle (see Figure 8.11 for more details). Reproduced with permission from [11].

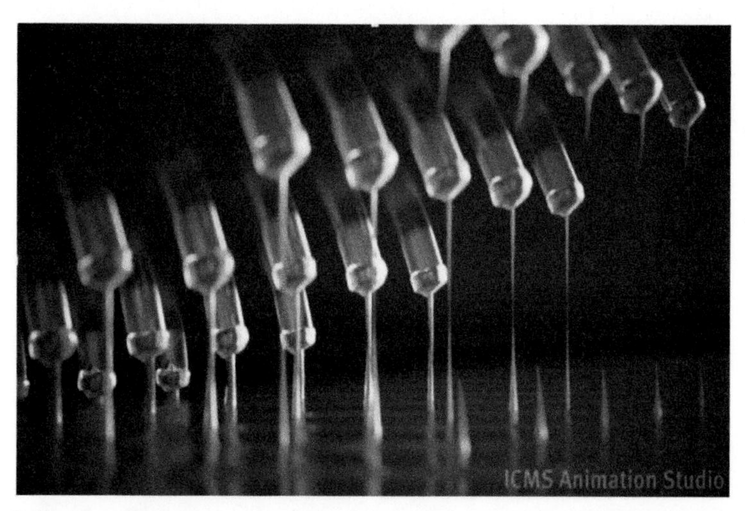

Figure 8.13 In [12], yet another manufacturing method for MAC is presented. In a "roll-pulling method," a glass substrate is covered with a film of uncured polymer (mainly PDMS) with embedded magnetic microparticles. Then, a roll covered with micropillars rolls over the precursor film and as the pillars touch the precursor film and move upward, small fibers are drawn out of the film, which are subsequently cured to stabilize them. The image shows the fibers being pulled out of the film. Copyright ICMS Animation Studio 2016.

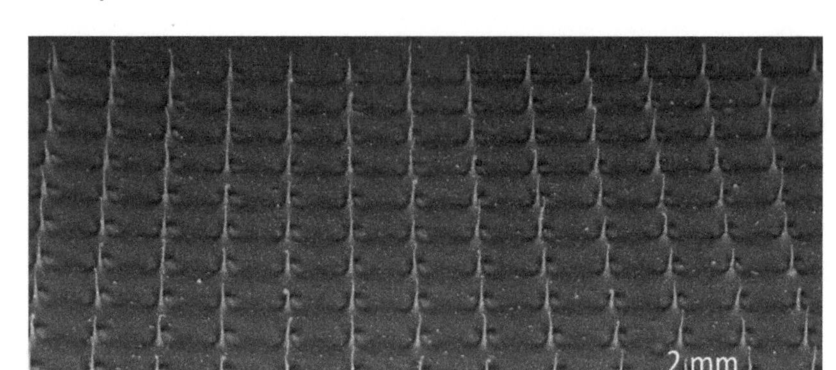

Figure 8.14 The roll-pulling method of Figure 8.13 makes it possible to create large areas of well-defined MAC. This SEM image shows an example. Reproduced with permission from [12].

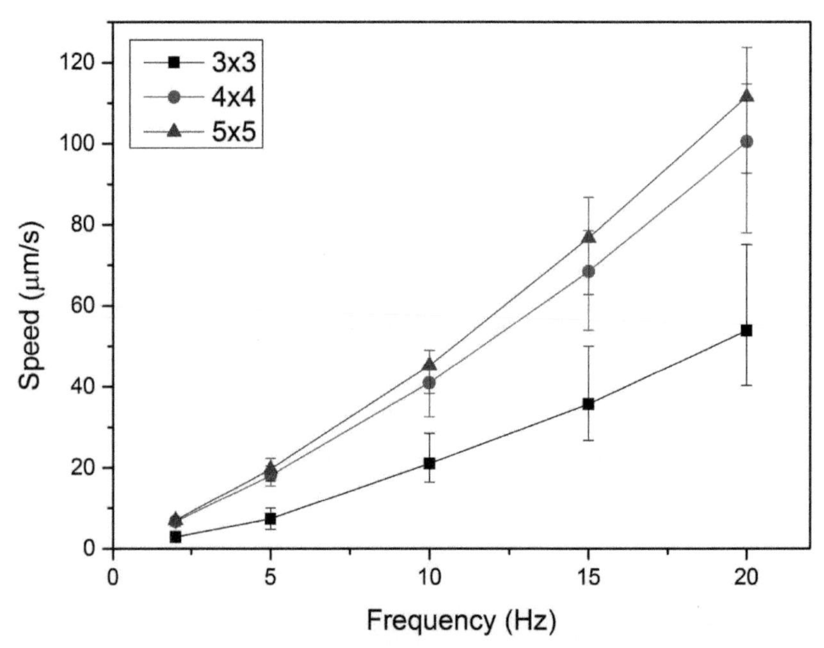

Figure 8.15 The roll-pulled MAC were integrated in a microfluidic channel in arrays of 3 × 3, 4 × 4, and 5 × 5 cilia. The cilia were rotated along a tilted cone using the setup of Figure 8.10 and the mean flow speed generated was quantified using PTV. This figure shows the flow speeds of water observed versus the actuation frequency. Reproduced with permission from [12].

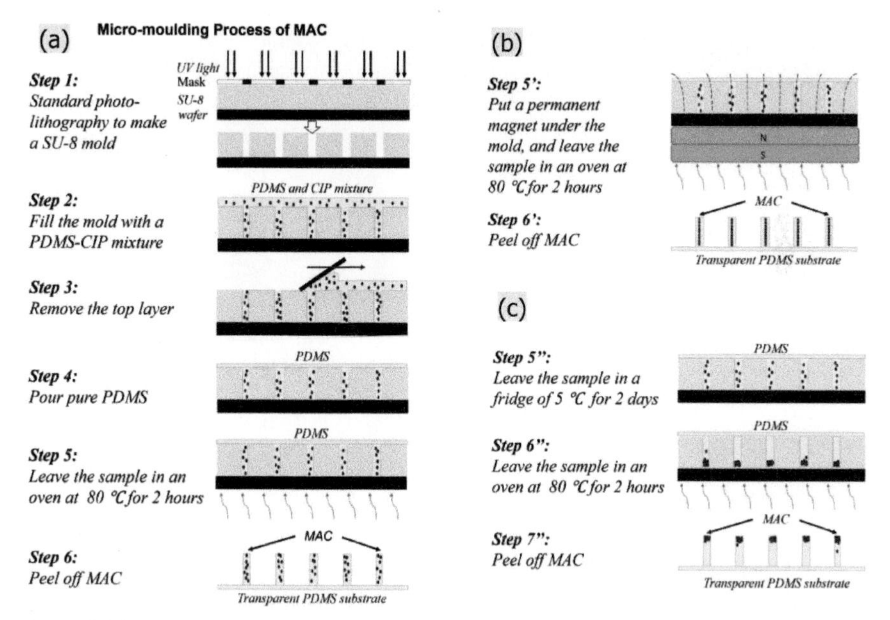

Figure 8.16 In [13], a micromolding method to fabricate MAC is presented. Applying a magnetic field during the molding step allows controlling the distribution of magnetic particles within the cilia. This figure shows a schematic of the micromolding process to fabricate MAC. (a) Standard process to fabricate MAC with a random magnetic particle distribution (standard MAC). (b) Last two steps to create MAC with a linearly aligned magnetic particle distribution (LAP MAC) employing a permanent magnet with a remnant flux density of 1.3 T to guide the distribution of magnetic particles within cilia. (c) Last three steps to make MAC with a concentrated particle distribution in cilia tips (CP MAC) using gravity induced sedimentation of magnetic particles. Illustrations are not to scale. Reproduced with permission from [13].

increased viscous drag, like discussed earlier. Figure 8.12 shows the flow characterization (water) and the dynamics of MAC made with magnetic fiber drawing. Measured flow speed is shown along with the theoretical prediction $f\alpha(\sin\theta\sin^2\Phi)$ at different actuation frequencies f, in which θ is the cone tilt angle and Φ is the cone opening angle measured on the short axis of the ellipsoidal shaped cone. Change of the aspect ratio of the ellipsoidal cone at different frequencies is shown in the embedded graph. Snapshots at various frequencies show the change in the trace of cilia during one actuation cycle. Similar to the behavior shown in Figure 8.8, the velocity has a maximum; however, it occurs at higher actuation frequency, and the amplitude is much higher, namely, 70 μm/s.

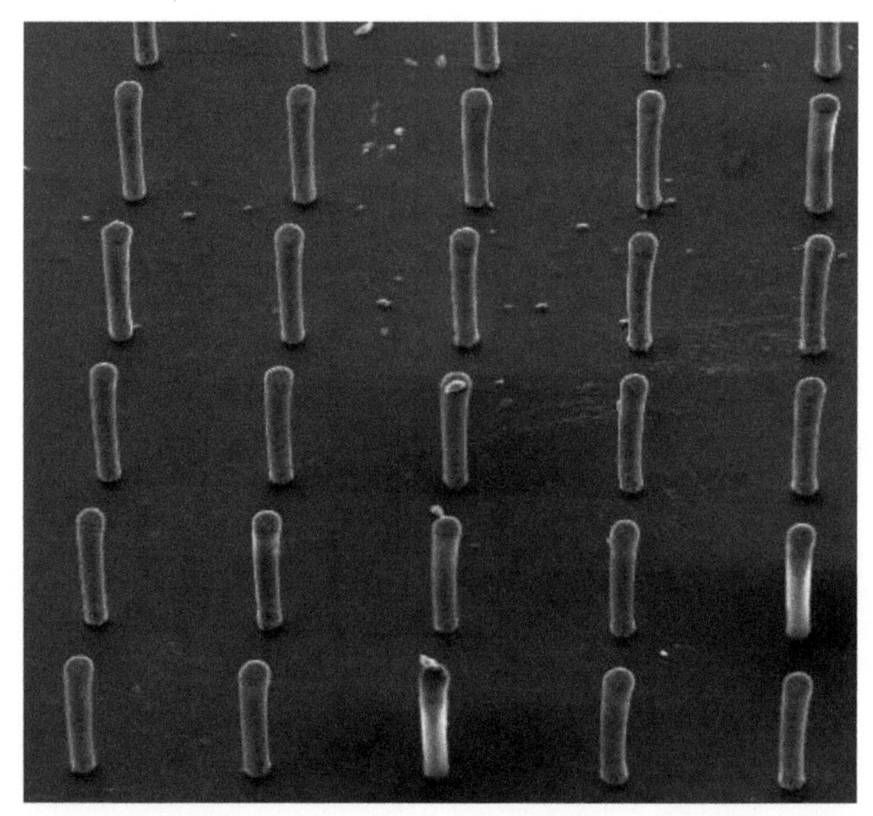

Figure 8.17 Electron micrograph of the MAC made with the micromolding methods shown in Figure 8.16, taken at a 30° angle. The cilia have a diameter, a length, and a pitch of 50, 350, and 350 μm, respectively. Reproduced with permission from [13].

To be able to produce MAC very efficiently and on large surface areas, we developed the "roll-pulling" process, illustrated in Figure 8.13 and reported in [12]. In this process, a glass substrate is covered with a film of uncured polymer (mainly PDMS) with embedded magnetic microparticles. Then, a roll decorated with micropillars rolls over the precursor film that is translated underneath, and as the pillars touch the precursor film and move upward, small fibers are drawn out of the film, which are subsequently cured to stabilize them. The image shows the fibers being pulled out of the film. The length of these artificial cilia is typically 250 μm. The scanning electron microscope (SEM) image of Figure 8.14 shows that the roll-pulling method indeed makes

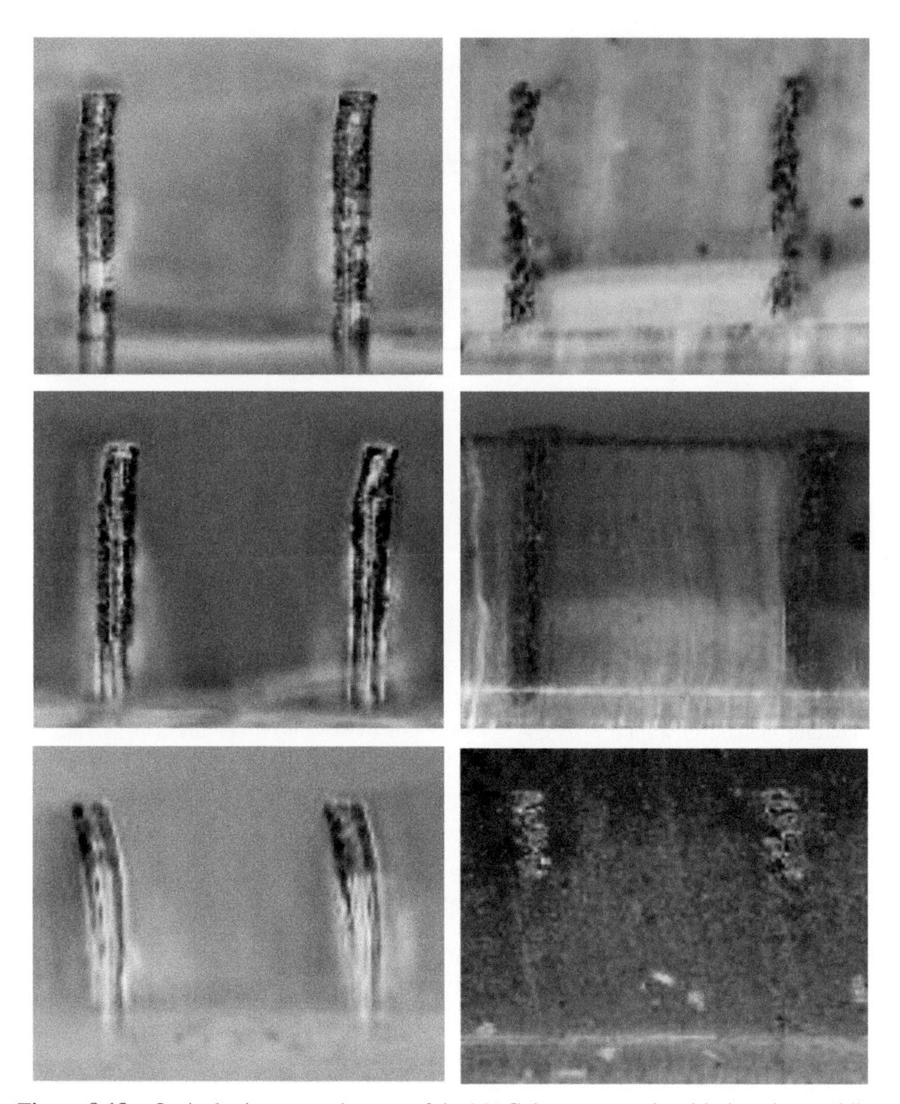

Figure 8.18 Optical microscopy images of the MAC that were made with the micromolding method of Figure 8.16. Top: the standard MAC (with magnetic particles distributed randomly within the cilia), middle: the LAP MAC (with the magnetic particles lined up in the cilia length direction), and bottom: the CP MAC (with the magnetic particles concentrated in the cilia tips), in both side view (left images) and cross-sectional view (right images). The cross sections were made by first embedding the cilia fully in PDMS, curing, and then cross sectioning. All scale bars are 350 μm. Reproduced with permission from [13].

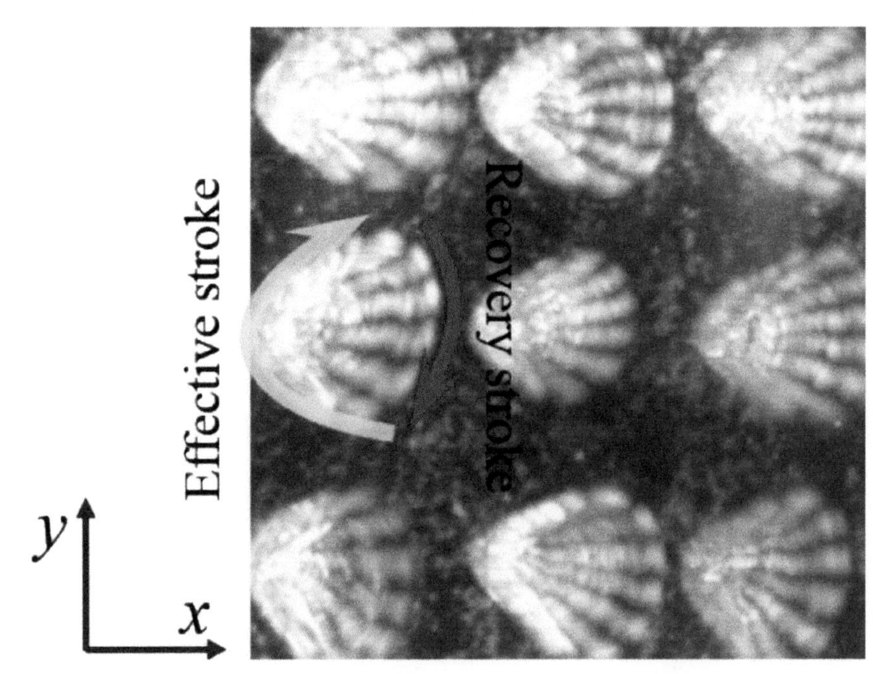

Figure 8.19 Top-view image of the motion of the rotating LAP MAC (made using the micromolding method of Figure 8.16 and actuated using the setup of Figure 8.10) in air, showing the tilted cone rotation, composed of 25 frames in one actuation cycle at 40 Hz. The scale bar is 350 μm. Reproduced with permission from [13].

it possible to create large areas of slender MAC. After fabrication, the roll-pulled MAC were integrated in a microfluidic channel in arrays of 3×3, 4×4, and 5×5 cilia. The cilia were rotated along a tilted cone using the setup of Figure 8.10 and the mean flow speed generated was quantified using particle tracking velocimetry (PTV). Figure 8.15 shows the flow speeds of water observed versus the actuation frequency. Not surprisingly, the more the cilia are integrated, the higher the generated flow speed is. Also, the speed goes up with rotation frequency, and has not yet reached the critical speed at which the viscous drag will start counteracting the cilia motion in Figure 8.15. The largest fluid velocities obtained using these artificial cilia and this particular setup were over 100 μm/s.

The roll-pulling process is indeed an efficient process to fabricate MAC on large surface areas. However, a drawback of the process is that the control of the precise size and shape of the cilia is difficult, since it is dictated by the pulling process and the rheological properties of the precursor film.

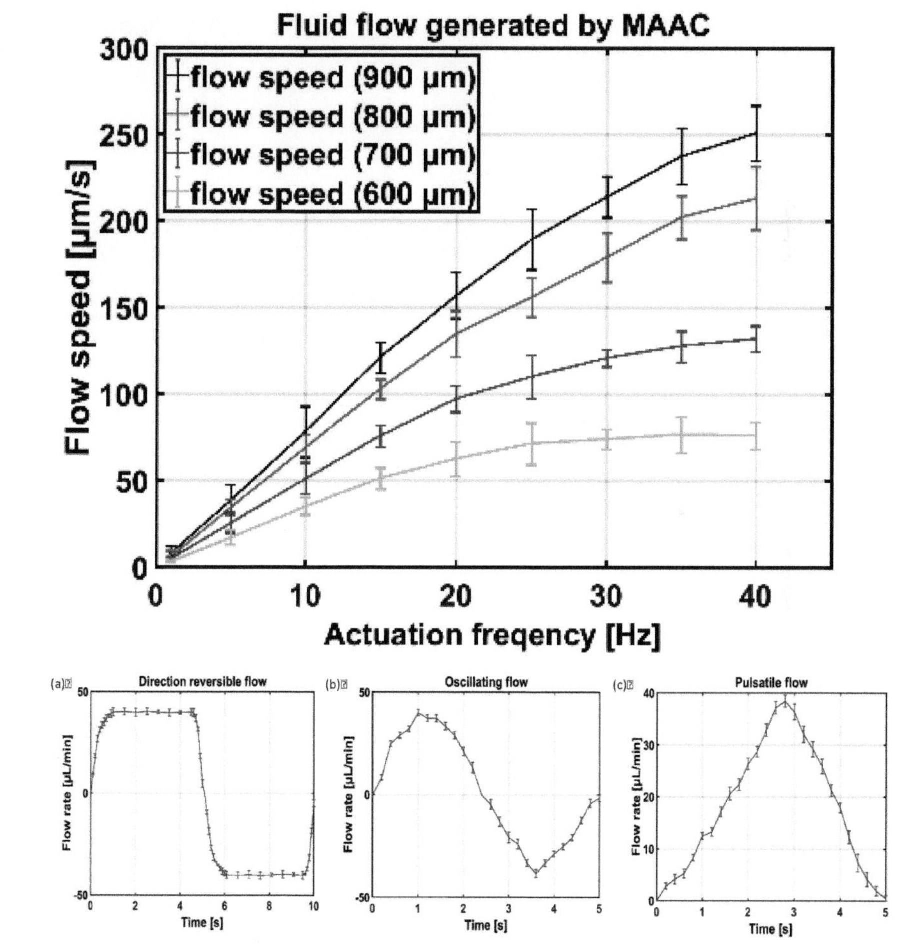

Figure 8.20 Fluid flow generation characterization of the MAC ("LAP MAC") made using the micromolding method of Figure 8.16. The setup of Figure 8.10 was used. The upper graph shows the mean water speed generated in microchannels of different heights, as a function of rotation frequency of the cilia. The speed can go up to 250 μm/s. By controlling the rotation of the magnetic field, shown in Figure 8.10, in time, specific time-dependent flow velocity profiles could be generated as shown in the bottom figures: direction reversible flow, oscillating flow, and pulsatile flow. Reproduced with permission from [13].

For control of cilia shape and magnetic particle distribution within the cilia, we introduced a micromolding method in [13]. Figure 8.16 illustrates the process. Applying a magnetic field during the molding step allows controlling the distribution of magnetic particles within the cilia. Figure 8.16a illustrates

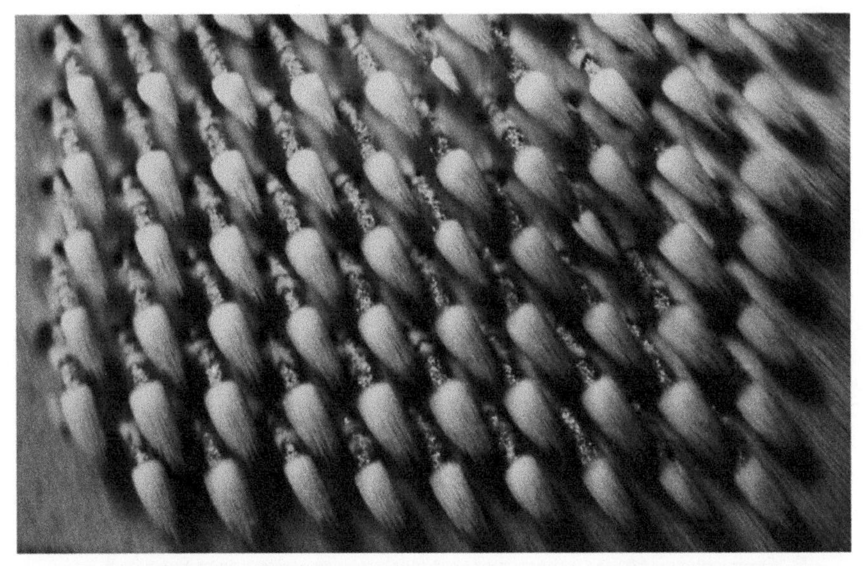

Figure 8.21 Optical micrograph of the MAC made using micromolding (Figure 8.16). These cilia are 350 μm in length. In this experiment, additional magnetic particles were added in the fluid, creating elongated magnetic clusters attached to the cilia tips. Copyright S. Zhang 2018.

our standard process to fabricate MAC with a random magnetic particle distribution (standard MAC). The last two steps shown in Figure 8.16b enable creating MAC with a linearly aligned magnetic particle distribution (LAP MAC) employing a permanent magnet with a remnant flux density of 1.3 T to guide the distribution of magnetic particles within cilia. In the process shown in Figure 8.16c, the last three steps serve to make MAC with a concentrated particle distribution in cilia tips (CP MAC) using gravity-induced sedimentation of magnetic particles. Measurement of the magnetic moment of the cilia showed that the LAP MAC have substantially larger magnetic moments than the other cilia, making it easier to actuate them. In Figure 8.17, an electron micrograph is shown for the MAC made with the micromolding method of Figure 8.16, taken at a 30° angle. The cilia have a diameter, a length, and a pitch of 50, 350, and 350 μm, respectively. Optical microscopy images of the three different types of MAC that were made with the micromolding method are depicted in Figure 8.18. The difference in distribution of magnetic particles within the different cilia can be appreciated from these images. The LAP MAC were integrated in microfluidic chips, and actuated using the setup of Figure 8.10. Figure 8.19 shows top-view

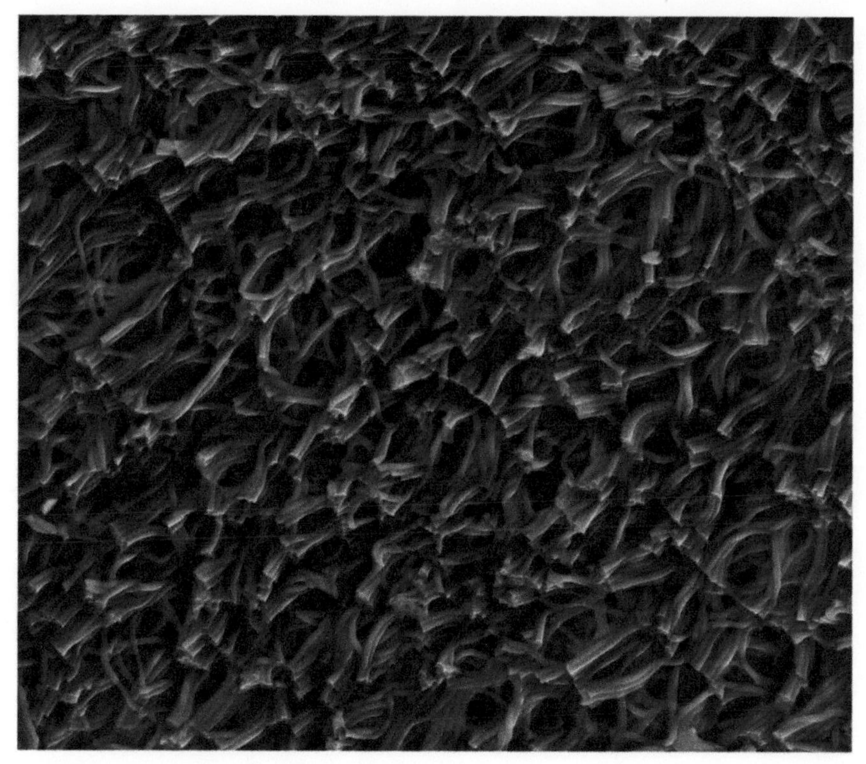

Figure 8.22 Artificial cilia using a micromolding method similar to that depicted in Figure 8.16, but now a "track etched membrane" was used as a mold. The cilia are 22 μm long and about 3 μm thick, and they form a dense cilia carpet. The image was taken using SEM, and false colors are used. Copyright Y. Wang 2018.

images of the motion of the rotating artificial cilia in air, showing the tilted cone rotation, composed of 25 frames in one actuation cycle at 40 Hz. The scale bar is 350 μm. Figure 8.20 demonstrates the fluid flow generation capabilities of the MAC ("LAP MAC"). The upper graph shows the mean water speed generated in microchannels of different heights, as a function of rotation frequency of the cilia. The fluid speed can go up to 250 μm/s. By controlling the rotation of the magnetic field, using the setup shown in Figure 8.10, in time, specific time-dependent flow velocity profiles can be generated as shown in the bottom figures: direction reversible flow, oscillating flow, and pulsatile flow can be induced. This is a unique feature that is very hard to achieve with conventional micropumps such as syringe pumps that are often used to drive fluids in microfluidics devices. An optical micrograph

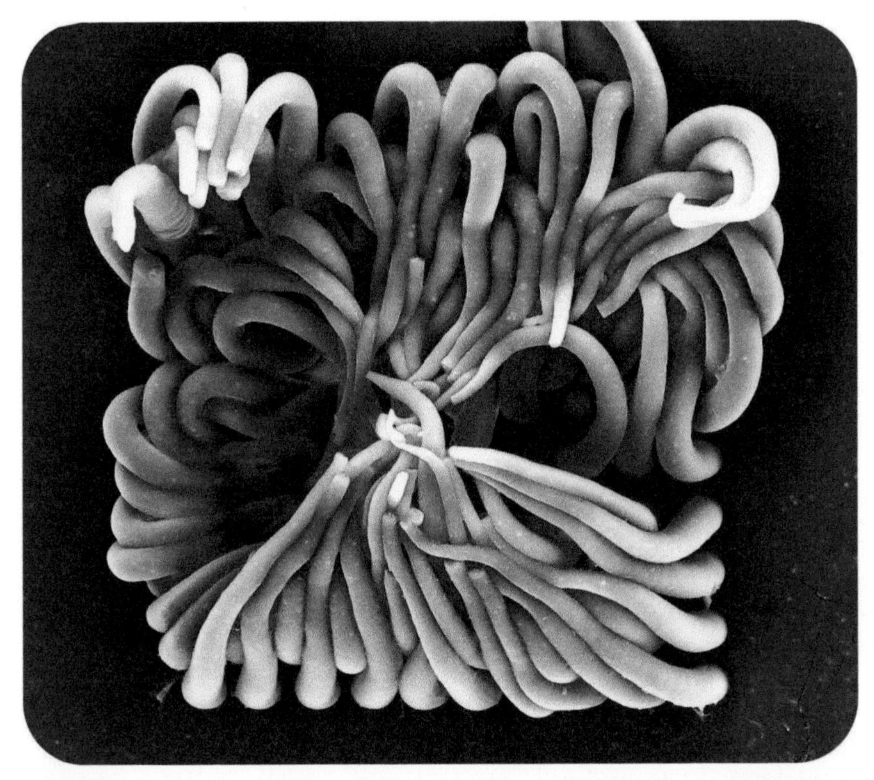

Figure 8.23 Artificial cilia. To produce these, first, a glass mold was produced using femtosecond laser writing and hydrofluoric acid etching [29]. PDMS (with dispersed magnetic nanoparticles) was poured in this mold, cured, and carefully released. The cilia are 100 μm long and about 10 μm thick. The image was taken using SEM, and false colors are used. Reproduced with permission from [28].

of the MAC made using micromolding is depicted in Figure 8.21. These cilia are 350 μm in length. In this experiment, additional magnetic particles were added in the fluid, creating elongated magnetic clusters attached to the cilia tips.

Finally, we show here two more artificial cilia created in our lab. In Figure 8.22, artificial cilia are shown that were fabricated using a micromolding method similar to that depicted in Figure 8.16, but now a "polycarbonate track etched membrane" was used as a mold. The cilia are 22 μm long and about 3 μm thick, and they form a dense cilia carpet. Figure 8.23 shows a fascinating image of artificial cilia that were fabricated by, first, producing a glass mold using femtosecond laser writing and hydrofluoric acid etching [29]. PDMS (with dispersed magnetic nanoparticles) was poured in this mold, cured, and carefully released. The cilia are 100 μm long and about 10 μm thick.

References

[1] den Toonder JMJ, and Onck PR. Microfluidic manipulation with artificial/bioinspired cilia. *Trends in Biotechnology*, 31:85–91, 2013.
[2] den Toonder JMJ, and Onck PR. (eds). *Artificial Cilia*. Cambridge: RSC Publishing, 2013.
[3] den Toonder JMJ, Bos FM, Broer DJ, et al. Artificial cilia for active micro-fluidic mixing. *Lab on a Chip*, 8:533–541, 2008.
[4] Evans BA, Shields AR, Carroll RL, et al. Magnetically actuated nanorod arrays as biomimetic cilia. *Nano Lett*. 7:1428–1434, 2007.
[5] Shields AR, Fiser BL, Evans BA, et al. Biomimetic cilia arrays generate simultaneous pumping and mixing regimes. *Proc. Natl. Acad. Sci. U.S.A.*, 107:15670–15675, 2010.
[6] Fahrni F, Prins MWJ, van IJzendoorn LJ, Micro-fluidic actuation using magnetic artificial cilia. *Lab on a Chip*, 9:3413–3421, 2009.
[7] Vilfan M, Potonik A, Kavi B, et al. Self-assembled artificial cilia. *Proc. Natl. Acad. Sci. U.S.A.*, 107:1844–1847, 2010.
[8] Hussong J, Schorr N, Belardi J, et al. Experimental investigation of the flow induced by artificial cilia. *Lab on a Chip*, 11:2017, 2011.
[9] Babataheri A, Roper M, Fermigier M, et al. Tethered fleximags as artificial cilia. *J. Fluid Mech.*, 678:5–13, 2011.
[10] Wang Y, Gao Y, Wyss HM, et al. Out of the cleanroom, self-assembled magnetic artificial cilia. *Lab on a Chip*, 13(17):3360–3366, 2013.
[11] Wang Y, Gao Y, Wyss HM, et al. Artificial cilia fabricated using magnetic fiber drawing generate substantial fluid flow. *Microfluidics and Nanofluidics*, 18:167–174, 2015.
[12] Wang Y, Cardinaels R, Anderson PD, et al. A continuous roll-pulling approach for the fabrication of magnetic artificial cilia with microfluidic pumping capability. *Lab on a Chip*, 16:2277–2286, 2016.
[13] Zhang S, Wang Y, Lavrijsen R, et al. Versatile microfluidic flow generated by moulded magnetic artificial cilia. *Sensors and Actuators B*, 263:614–624, 2018.
[14] Khaderi SN, Hussong J, Westerweel J, et al. Fluid propulsion using magnetically-actuated artificial cilia–experiments and simulations. *RSC Advances*, 3:12735, 2012.

[15] Khaderi SN, Craus CB, Hussong J, et al. Magnetically actuated artificial cilia for microfluidic propulsion. *Lab on a Chip*, 11:2002, 2011.
[16] Khaderi SN, den Toonder JMJ, and Onck PR. Magnetic artificial cilia for microfluidic propulsion. *Advances in Applied Mechanics*, 48:1, 2016.
[17] Gorissen B, de Volder M, and Reynaerts D. Pneumatically-actuated artificial cilia array for biomimetic fluid propulsion. *Lab on a Chip*, 15(22):4348–4355, 2015.
[18] Oh K, Smith B, Devasia S, et al. Characterization of mixing performance for bio-mimetic silicone cilia. *Microfluid. Nanofluidics*, 9(4–5):645–655, 2010.
[19] Oh K, Chung J, Devasia S, et al. Bio-mimetic silicone cilia for microfluidic manipulation. *Lab on a Chip*, 9:1561–1566, 2009.
[20] van Oosten CL, Bastiaansen CWM, and Broer DJ. Printed artificial cilia from liquid-crystal network actuators modularly driven by light. *Nature Mater.*, 8:677–682, 2009.
[21] Zarzar LD, Kim P, and Aizenberg J. Bio-inspired design of submerged hydrogel-actuated polymer microstructures operating in response to pH. *Adv. Mater.*, 23:1442–1446, 2011.
[22] Khademolhosseini F, and Chiao, M. Fabrication and patterning of magnetic polymer micropillar structures using a dry-nanoparticle embedding technique. *J. Microelectromechanical Syst.*, 22(1):131–139, 2013.
[23] Glazer PJ, Leuven J, An H, et al. Multi−Stimuli Responsive Hydrogel Cilia. *Advanced Functional Materials*, 23(23):2964–2970, 2013.
[24] Chen C-Y, Chen C-Y, Lin C-Y, et al. Magnetically actuated artificial cilia for optimum mixing performance in microfluidics. *Lab on a Chip*, 13:2834–2839, 2013.
[25] Baltussen MGHM, Anderson PD, Bos FM, et al. Inertial flow effects in a micro-mixer based on artificial cilia. *Lab on a Chip*, 9(16):2326–2331, 2009.
[26] Belardi J, Schorr N, Prucker O, et al. Artificial cilia: Generation of magnetic actuators in microfluidic systems. *Advanced Funcional Materials*, 21(17):3314–3320, 2011.
[27] Khaderi SN, Craus CB, Hussong J, et al. Magneticallyactuated artificial cilia for microfluidic propulsion. *Lab on a Chip*, 11:2002, 2011.
[28] Wang Y. Artificial Life - the winner of the Art in Science Award presented at MicroTAS 2013. *Lab on a Chip*, 14, 2014.
[29] Bellouard Y, Said A, Dugan M, et al. Fabrication of high-aspect ratio, micro-fluidic channels and tunnels using femtosecond laser pulses and chemical etching. *Optics Express*, 12:2120–2129, 2004.

PART III

Theoretical and Computer Modelling

9

Numerical Simulations of Fluid Transport by Magnetically Actuated Artificial Cilia

Syed Khaderi[1], Jaap M. J. den Toonder[2] and Patrick Onck[3]

[1]Department of Mechanical and Aerospace Engineering, Indian Institute of Technology Hyderabad, Kandi, Hyderabad 502285, India
[2]Department of Mechanical Engineering, Eindhoven University of Technology, Eindhoven, 5600 MB, The Netherlands
[3]Zernike Institute for Advanced Materials, University of Groningen, Groningen, 9747 AG, The Netherlands
E-mail: snk@iith.ac.in; j.m.j.d.toonder@tue.nl; p.r.onck@rug.nl

The working mechanism of cilia comprises an effective and a recovery stroke that are different in nature. This unique asymmetric motion of the cilia is an essential feature to generate net fluid flow at low Reynolds numbers. Inspired by this asymmetric mechanism, thin-film artificial cilia have been fabricated to propel fluids in the microchannels of lab-on-a-chip biosensors. The artificial cilia are polymer films embedded with magnetic nanoparticles and are actuated by an external magnetic field, mimicking the beating of natural cilia. In this chapter, we describe the basic results obtained from an extensive numerical analysis of generating fluid transport using magnetically driven artificial cilia.

9.1 Introduction

Lab-on-a-chip (LOC) technology aims at performing analyses of biological samples (such as blood and urine), conventionally performed in a clinical lab at a medical center, on a small chip. The LOC analyses range from simple chemical tests on biological samples to sophisticated DNA and cell

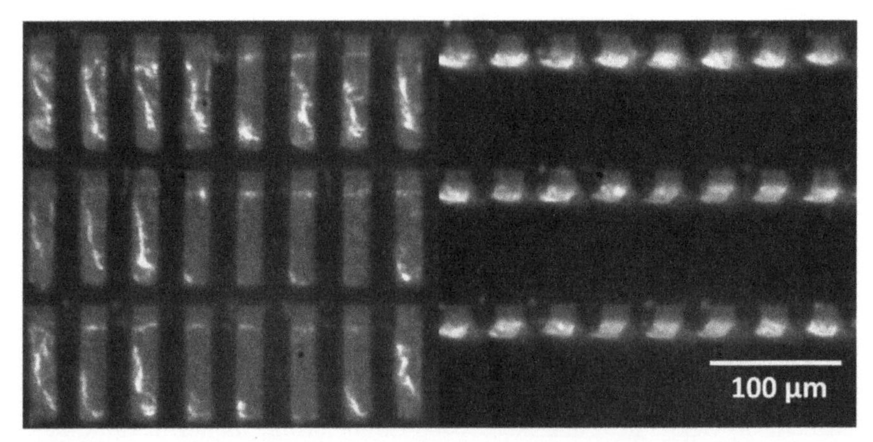

Figure 9.1 Top view of a surface covered with magnetic artificial cilia fabricated by the group of Prof. Ruhe at the University of Freiburg. These magnetic artificial cilia are polymer thin films manufactured using photolithography techniques and are embedded with super-paramagnetic nanoparticles. Left: Artificial cilia in the inactivated state. Right: Artificial cilia under the influence of a magnetic field showing an out-of-plane deformation. Reproduced with permission from [1].

analysis. Since the surface-to-volume ratio is high in an LOC, physical phenomena associated with surfaces (e.g., surface tension and electrokinetics) gain importance. In addition, as the length scales involved are small (typically less than a millimeter), the viscous forces in the fluid dominate over the inertial forces leading to laminar flow profiles in typical LOCs. While some of these properties are beneficial for certain applications, for others, they can be detrimental. For instance, the fabrication of devices to perform individual operations on an LOC cannot always be done by simply downscaling conventional methodologies. Mixing of an analyte with another fluid is difficult to achieve in a microfluidic device due to the laminar nature of flow at these length scales. Another challenge is the pumping of fluids through the microchannels and testing chambers on an LOC. In some applications, a local control of the flow is necessary, which calls for a localized pumping system that can be embedded into a microchannel.

One example of micron-scale fluid propulsion is the flow of mucus out of our lungs by the motion of hair-like microscopic structures, called cilia. Their working principle comprises a forward and a recovery stroke that are spatially distinct so that more fluid is propelled in the direction of the forward stroke than that of the return stroke. This asymmetric motion of the cilia is a necessary prerequisite to generate a net fluid flow due to the dominance

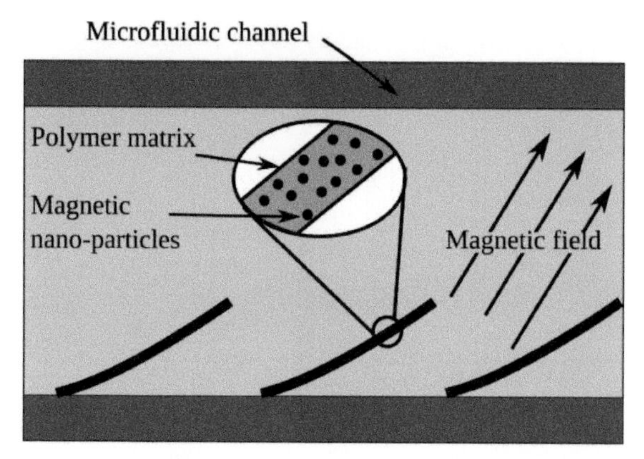

Figure 9.2 Side-view schematic picture of artificial cilia attached to the surface of a microchannel. The artificial cilia consist of a polymer matrix with embedded magnetic nano-particles. The cilia exhibit an asymmetric motion when a tuned external magnetic field is applied. Reproduced with permission from [2].

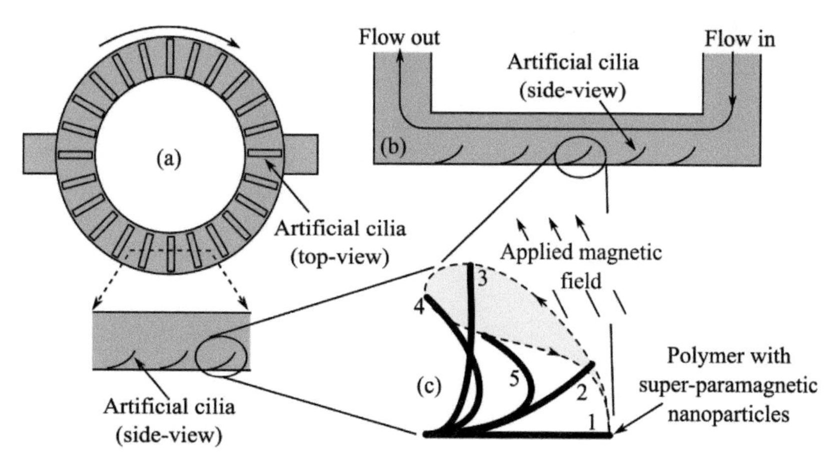

Figure 9.3 Two possible applications of artificial cilia in microfluidics: (a) closed-loop channel and (b) open-loop channel. In the closed-loop channel, the cilia can be used to propel the fluid inside a circular channel for a well-defined period of time. The closed-loop channel can, for example, be used to perform a PCR. In the open-loop channel, the cilia propel the fluid from the inlet of the channel to the outlet. (c) Schematic representation of the typical motion of a magnetically actuated artificial cilium during its beat cycle. The shaded region bounded by a dashed curve represents the area swept by the cilium. The direction of motion of the cilium is shown using the arrow on the dashed curve. The effective stroke is represented by instances 1–3, and the recovery stroke by instances 4 and 5. Reproduced with permission from [1].

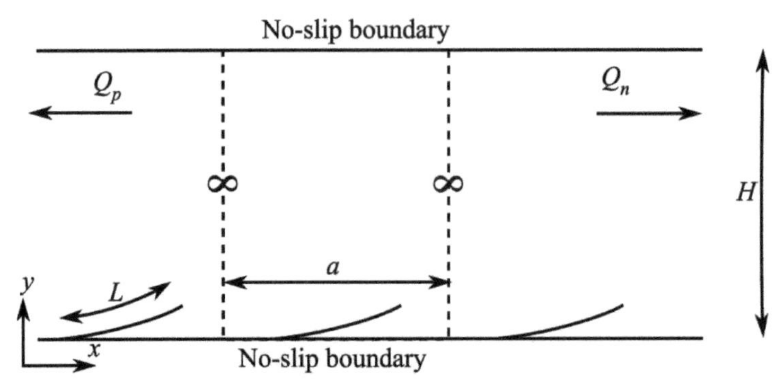

Figure 9.4 Schematic of the microfluidic channel with height H used to perform the simulations. Cilia of length L are arranged with an inter-ciliary spacing a. A periodic unit cell is used for the analysis, which is shown using the dashed lines. Reproduced with permission from [2].

of viscous effects over inertial effects, a consequence of the small length scales and low velocities. Inspired by this mechanism, thin-film artificial cilia have been fabricated [see Figure 9.1(left)]. When no magnetic field is applied, the cilia are oriented parallel to the surface of the substrate. When a magnetic field is applied that has a vectorial component out of the plane of the surface, the cilia deform and bend upward [Figure 9.1(right)]. In this work, we have developed a computational Euler–Lagrange finite-element approach that accounts for the simultaneous solution of the solid-mechanics, fluid-dynamics, and magneto-statics equations (see Figure 9.2 for a side-view schematic of the computational domain) [1–9]. The coupling of these three domains of engineering physics allows for the design of an optimal magnetic field rotation that maximizes the net fluid flow generated by the cilia. Further details of the numerical methodology can be found in [2].

Two typical channel configurations that are encountered in LOC applications are depicted in Figure 9.3: (i) a closed-loop channel (see Figure 9.3a) and (ii) an open-loop channel (see Figure 9.3b). In the closed-loop channel, fluid is initially pumped in by an external device and then it is propelled around by the artificial cilia. Closed-loop channels are used, for instance, to perform polymerase chain reaction (PCR). In an open-loop channel, on the other hand, we have well-defined inlet and outlet points for the fluid to enter and leave the channel. The fluid is propelled by an array of artificial cilia inside the channel. For the closed-loop channel, we assume the radius of the loop to be much larger than the height of the channel and the spacing between the cilia, so that we can analyze a straight, infinitely long channel with equally

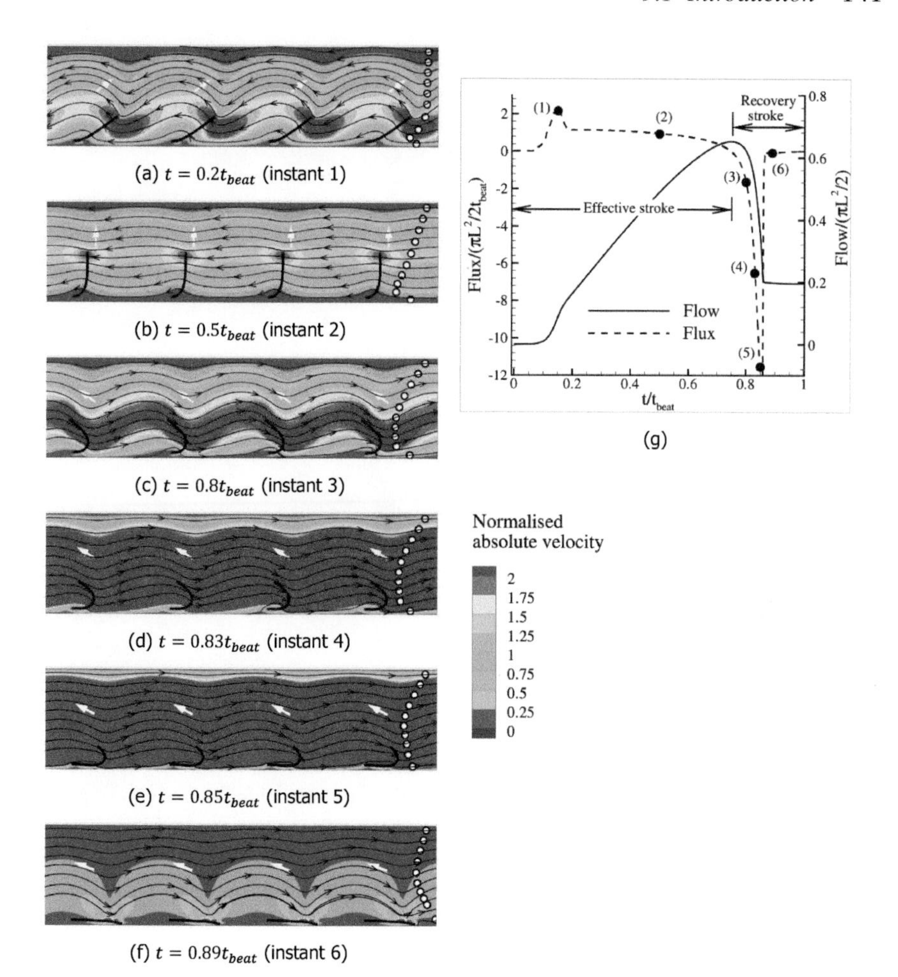

Figure 9.5 Results of finite element simulations of low Reynolds number fluid flow created by magnetic artificial cilia that are subjected to a rotating magnetic field. (a)–(f) Contours of absolute velocity (normalized with L/t_{beat}) at different time instants. The direction of the velocity is given by the streamlines and the white arrows represent the magnetic field at the respective time instances. The circles represent fluid particles. Four unit cells are shown for clarity. Note that the white circles move to the left and to the right during the effective and recovery stroke, respectively. However, at the end of one beat cycle, they are displaced to the left, indicating a positive net fluid flow during one beat cycle. (g) Instantaneous flux and accumulated flow as a function of time. The time instances corresponding to (a)–(f) are duly marked. During the effective stroke, the flux is positive and the flow increases with time. During the recovery stroke, the flux is negative and the flow decreases. However, at the end of one cycle of operation, the net flow is positive, indicating a net fluid flow in the direction of the effective stroke. Reproduced with permission from [2].

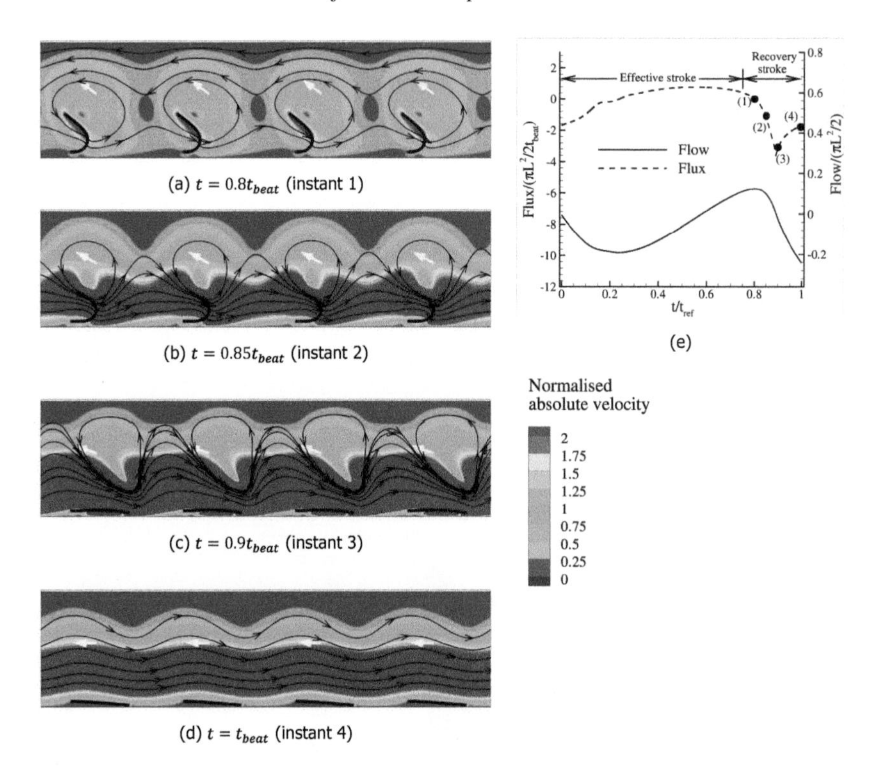

Figure 9.6 Results of finite-element simulations of moderate Reynolds number fluid flow created by magnetic artificial cilia that are subjected to a rotating magnetic field. (a)–(d) Contours of absolute velocity (normalized with L/t_{beat}) at different time instants. The direction of the velocity is given by the streamlines and the white arrows represent the magnetic field at the respective time instances. Four unit cells are shown for clarity. The flow contours during the effective stroke are similar to Figure 9.5, and hence only snapshots during the recovery stroke are shown. The footprint of inertia is clearly present when fluid flow continues to proceed in the direction of the recovery stroke (to the right) when the cilia are idle (see (c) and (d)). (e) Instantaneous flux (right axis) and flow (or accumulated flux, left axis) as a function of time with the instants (a)–(f) duly marked. The flux is negative for most of the time, leading to a negative net fluid flow (i.e., to the right). Reproduced with permission from [2].

spaced cilia. In that case, the analysis can be performed by using a periodic unit cell which contains only one cilium (see Figure 9.3). However, in the case of the open-loop channel, there is a no periodicity due to the presence of the inlet and outlet, and the analysis has to be performed with the channel containing multiple cilia. The cilia span the entire width of the channels for an optimal performance (see Figure 9.3), and the width of the channels is taken to be larger than the height. As a result, two-dimensional simulations

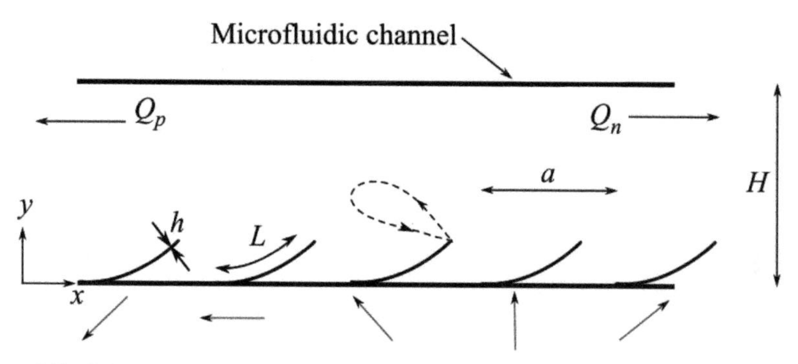

Figure 9.7 Schematic representation of the geometry used to analyze the flow due to out-of-phase motion of artificial cilia. We study an infinitely long microfluidic channel consisting of equal-sized cilia spaced a distance a apart. The variation of magnetic field in space at a given instance of time is shown using blue arrows. The magnetic field at each location rotates with a time period $2t_{beat}$. The magnetic field appears to travel to the right. Q_p and Q_n denote the flow in the direction of the effective and recovery stroke, respectively. Reproduced with permission from [3].

are sufficient for both channel geometries. In this article, we focus only on the closed-loop channel; for results on the open-loop channel, the reader is referred to [1].

First, we study the fluid propulsion created by a two-dimensional array of tapered super-paramagnetic artificial cilia, which are actuated by a uniform rotating magnetic field of magnitude B_0 and a time period t_{ref}. It is found that the time taken by the cilia for one beat cycle $t_{beat} = t_{ref}/2$. To perform the simulations, we take a unit cell containing one cilium (Figure 9.4). The top and bottom boundaries are no-slip boundaries and the left and right ends are periodic. The parameters are chosen so that fluid (viscous) forces are low compared to the elastic forces, but are large compared to the fluid inertia forces. i.e., the flow occurs at low Reynolds number.

When a rotating magnetic field is applied, the tapered artificial cilia exhibit an asymmetric motion as shown in Figure 9.3c. Figures 9.5a–f show snapshots of the cilia geometry, the magnitude of the fluid velocity (colors) and the flow direction (streamlines). The effective stroke is to the left (snapshots 1 and 2) and the recovery stroke is to the right (snapshots 3–6). The velocity of the cilia is higher during the recovery stroke when compared to the effective stroke. The cilia push the fluid to the left during the effective stroke and pull back the fluid during the recovery stroke. Nevertheless, due to the asymmetric motion, the fluid is transported in the direction of the effective stroke at this low Reynolds number (Figure 9.5). Next, we studied the effect

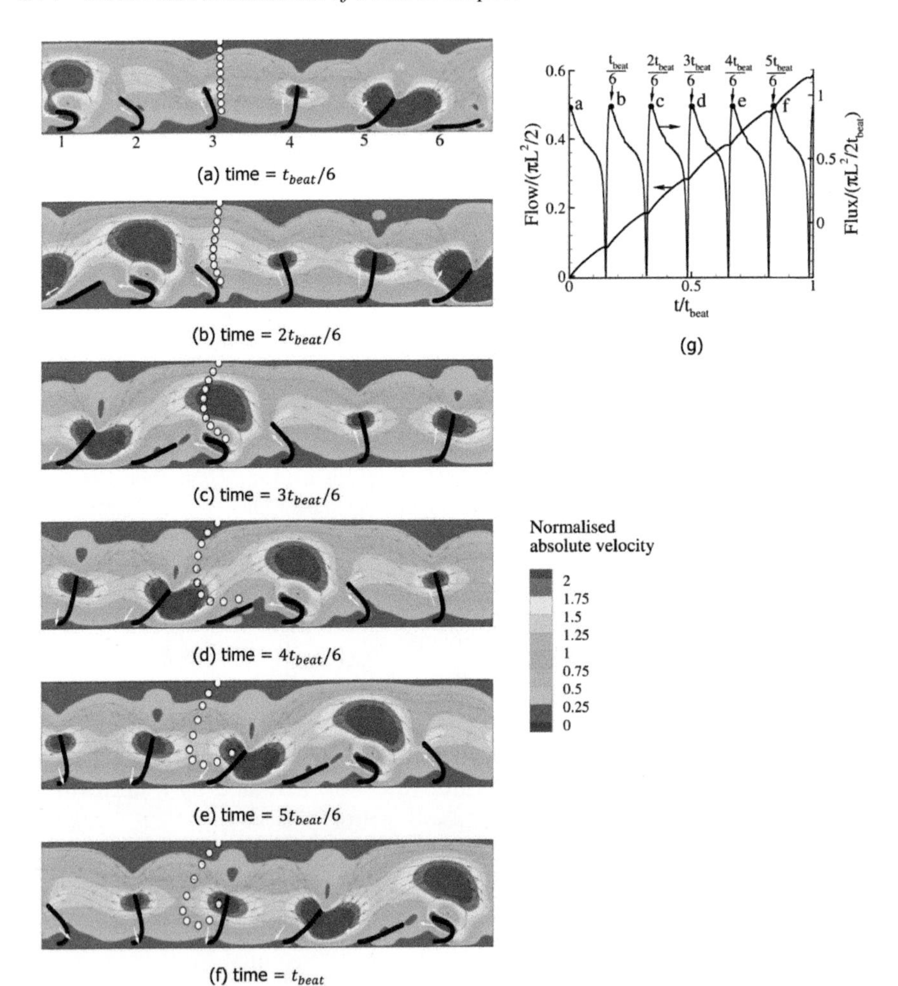

Figure 9.8 (a)–(f) Fluid transport due to out-of-phase motion of artificial cilia during a representative cycle. The magnetic wave travels to the right by six cilia in a beat cycle (time = t_{beat}). The contours represent the absolute velocity normalized with L/t_{beat}. The direction of the velocity is represented by streamlines. The applied magnetic field at each cilium is represented by the white arrows. The white circles represent fluid particles. The particles near the cilia show a fluctuating motion, while the particles away from the cilia continuously move to the left. (g) Instantaneous flux (right axis) and flow (or accumulated flux, left axis) as a function of time with the instants (a)–(f) duly marked. Unlike the case of uniform motion of cilia in Figure 9.5, the flux is positive for most of the time (i.e., unidirectional). Hence, the flow continuously increases with time with little fluctuations. Reproduced with permission from [3].

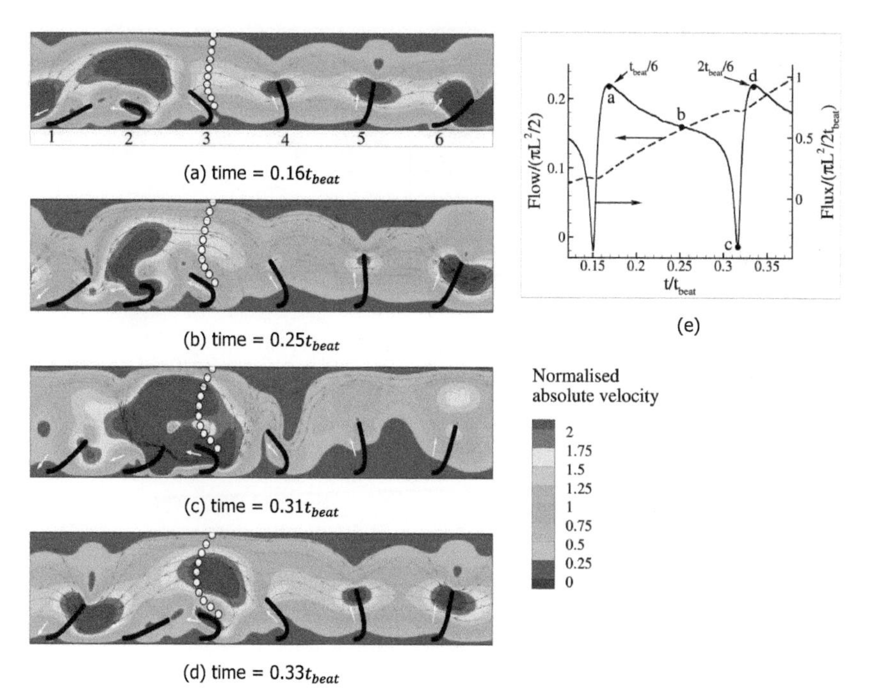

Figure 9.9 Flow analysis during the recovery stroke of the second cilium of Figure 9.8. (a)–(d) Snapshots for the out-of-phase motion of cilia between time instances of Figures 9.8b and c. The contours represent the absolute velocity normalized with L/t_{beat}. The direction of the velocity is represented by streamlines. The white circles represent fluid particles. The applied magnetic field at each cilium is represented by the white arrows. The second cilium is in the recovery stroke and causes a flow to the right. However, all the other cilia are in various phases of their effective stroke, which cause a flow to the left. The flow in the effective direction dominates the recovery flow due to the second cilium, leading to a net fluid flow in the direction of the effective stroke. (e) Instantaneous flux (right axis) and flow (left axis) as a function of time with the instances (a)–(d) duly marked. Although the second cilium is in the recovery stroke, the flux is always positive due to the flow created by the other cilia. Reproduced with permission from [3].

of the Reynolds number on the flow characteristics (Figure 9.6). Due to the enhanced fluid inertia at moderate Reynolds numbers, especially during the recovery stroke (instant 3, see Figure 9.6c), the cilia create a net fluid flow in the direction of the recovery stroke. For both values of the Reynolds number, due to the push and pull acting on the fluid, the fluid transported is pulsatile in nature.

We then proceed to analyze what happens when the applied magnetic field between adjacent cilia is slightly out of phase (see Figure 9.7; the analysis

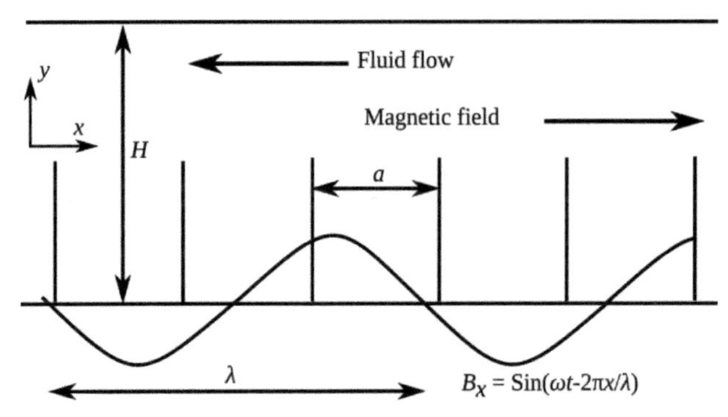

Figure 9.10 Fluid flow due to the out-of-phase oscillation of cilia, resembling a traveling wave. Schematic picture of the problem analyzed. We study an infinitely long channel of height H containing equally spaced cilia that are oriented perpendicular to the channel direction. The cilia are permanently magnetic, with the magnetization pointing along the cilia length. A magnetic field in the x-direction varies in space and time, whereas that in the y-direction is maintained constant. The magnetic field "wave" moves to the right. The resulting flow is to the left. Reproduced with permission from [4].

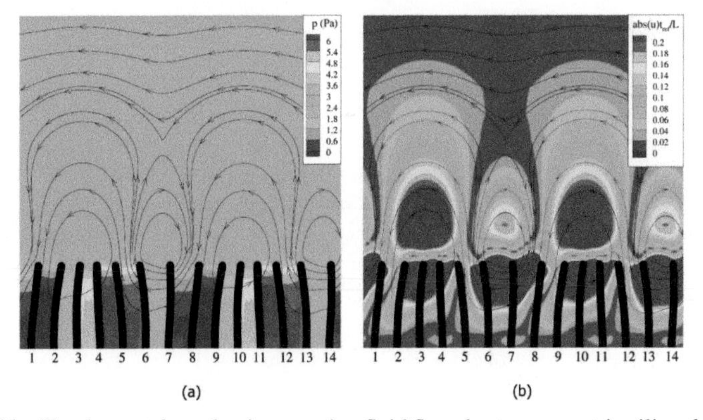

Figure 9.11 Fundamental mechanism causing fluid flow due to symmetric ciliary beating but with a finite out-of-phase motion representing a metachronal wave. (a) Contours of pressure and (b) contours of absolute velocity in the x-direction for $a/L = 2/7$ and $a/\lambda = 1/7$ (wave moving to the right) at time $t = 0.35t_{ref}$. At the instant depicted, cilia 2–4 and 9–11 move to the right, while cilia 6, 7 and 13, 14 move to the left. The other cilia are nearly stationary (zero velocity). Due to the instantaneous velocity of the cilia, high pressure (hp) and low pressure (lp) regions develop (red and blue regions). Fluid is squeezed out from the hp region and sucked in by the lp regions, as a result of which a series of counter-rotating vortices are formed in the channel. Since the distance between the hp and lp regions opposite to the wave direction is smaller, the pressure gradient is larger, so that the counterclockwise vortices are stronger. As a result, the velocity distribution has a dominant horizontal component to the left. Reproduced with permission from [4].

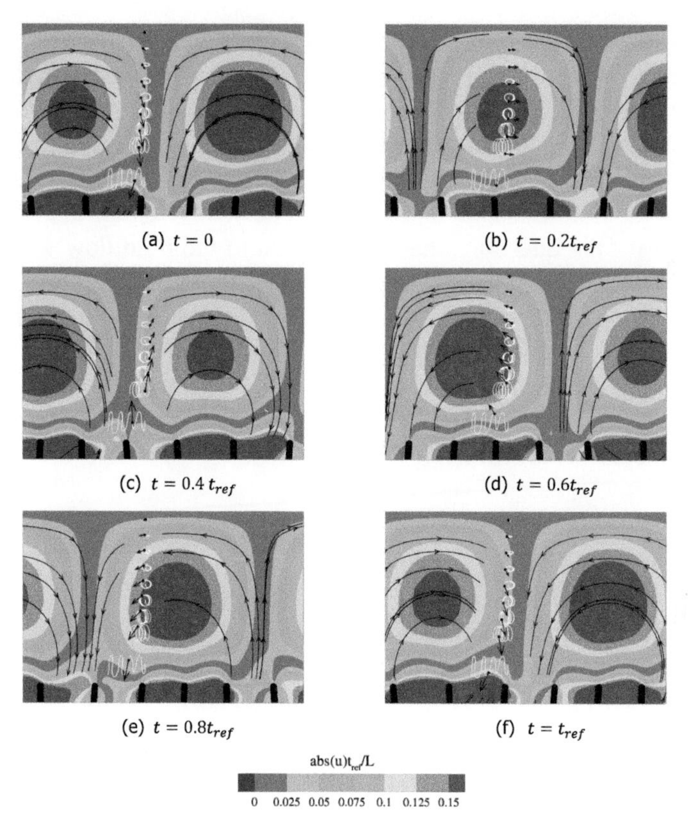

Figure 9.12 Motion of particles with time: the field-of-view is the region between the oscillating end of the cilia and the top boundary of the unit cell with the bottom-left corner at $(x, y) = (0.5L, 0.95L)$ and the top-right corner at $(x, y) = (1.55L, 2L)$. The velocity in the channel direction is larger in the direction opposite to the wave than in the direction of the wave. The white curves represent the trajectory of the particles and the black dots represent the particles. Let us focus our attention on the second particle from the bottom. At $t = 0$, the particle is between two vortices. The velocity of the particle is such that it moves downward. As time progresses, at $t = 0.2t_{ref}$, the position of the particle is such that it has a low velocity to the right due to the presence of the clockwise vortex. At time $t = 0.4t_{ref}$, the particle moves away from the influence of the clockwise vortex, toward the counter-clockwise vortex. Now the particle has a velocity such that it moves upward. At $t = 0.6t_{ref}$, when the particle is under the influence of the stronger counter-clockwise vortex, it has a higher velocity compared to the instance when the particle was under the influence of the less strong clockwise vortex (compare b and d). Therefore, the particle effectively moves to the left. Particles near the cilia move unidirectional and show larger displacement, whereas the particles near the top boundary do not show any displacement due to the no-slip boundary condition. Reproduced with permission from [4].

is performed at low Reynolds numbers). We see that a "wave" of ciliary motion travels to the right (see Figures 9.8 and 9.9). Note that at any particular instance during this metachronal wave, only one of the cilia will perform a recovery stroke. Hence, its effect to cause a negative flow is overcome by the other cilia that are in several stages of the effective stroke. This minimizes the fluctuations in the fluid transport and leads to a positive overall fluid flow.

We finally ask the question what will happen to the fluid flow when the cilia are oscillating symmetrically but are beating out of phase (Figures 9.10–9.12), which also leads to a traveling (metachronal) wave. It is found that the out-of-phase motion is sufficient to cause a net fluid flow. The deformation of the cilia leads to local high and low pressure fields that result in an effective pressure gradient that drives the flow. This fluid transport direction is opposite to that of the metachronal wave.

References

[1] Khaderi SN, Craus CB, Hussong J, et al. Magneticallyactuated artificial cilia for microfluidic propulsion. *Lab on a Chip*, 11:2002, 2011.

[2] Khaderi SN, den Toonder JMJ, and Onck PR. Magnetically actuated artificial cilia: The effect of fluid inertia. *Langmuir*, 28:7921, 2012.

[3] Khaderi SN, den Toonder JMJ, and Onck PR. Microfluidic propulsion by the metachronal beating of magnetic artificial cilia: a numerical analysis. *Journal of Fluid Mechanics*, 688:4, 2011.

[4] Khaderi SN, den Toonder JMJ, and Onck PR. Fluid flow due to collective nonreciprocal motion of symmetricallybeating artificial cilia. *Biomicrofluidics*, 6:014106, 2012.

[5] Khaderi SN, Baltussen MGHM, Anderson PD, et al. Nature-inspired microfluidic propulsion using magnetic actuation. *Physical Review E*, 79:046304, 2009.

[6] Khaderi SN, Baltussen MGHM, Anderson PD, et al. Breaking of symmetry in microfluidic propulsion driven by artificial cilia. *Physical Review E*, 82:027302, 2010.

[7] Khaderi SN, Hussong J, Westerweel J, et al. Fluid propulsion using magnetically-actuated artificial cilia–experiments and simulations. *RSC Advances*, 3:12735, 2012.

[8] Khaderi SN, den Toonder JMJ, and Onck PR. Magnetic artificial cilia for microfluidic propulsion. *Advances in Applied Mechanics*, 48:1, 2016.

[9] den Toonder JMJ, and Onck PR. Microfluidic manipulation with artificial/bioinspired cilia. *Trends in Biotechnology*, 81:85, 2013.

10

Modeling Gel-based Artificial Cilia

Olga Kuksenok[1], Ya Liu[2] and Anna C. Balazs[2,*]

[1]Materials Science and Engineering Department,
Clemson University, Clemson, SC 29634, USA
[2]Chemical Engineering Department, University of Pittsburgh,
Pittsburgh, PA 15261, USA
*Corresponding Author
E-mail: okuksen@clemson.edu; yal65@pitt.edu; balazs@pitt.edu

10.1 Introduction

Computational models that can predict the dynamic behavior of artificial cilia are vital for designing microfluidic chambers where these cilia could significantly expand the functionality of the device. For example, anchored to the floor of the microchambers, artificial cilia could mitigate the fouling of the apparatus or selectively trap cells and biomolecules flowing through the channels. There are, however, critical challenges in establishing design rules for creating such ciliated chambers. Namely, the design must involve an effective combination of materials components, as well as the appropriate stimuli that will permit the controllable actuation of the artificial cilia and the desired dynamic behavior. Moreover, the models used to simulate these systems must capture the complex interactions among the different materials, the applied stimuli, the surrounding flowing fluids, and chemicals or particulates in these fluids. Herein, we focus on computational approaches we recently developed to simulate the dynamic behavior of responsive cilial layers in complex fluids, and thus, provide guidelines for creating microfluidic systems with enhanced utility. What follows is not a comprehensive review, but rather highlights of our efforts in tackling the challenges noted above.

To illustrate how the appropriate combination of components can lead to controllable actuation, we describe our work on coupling chemo-responsive gels with oscillating chemical reactions to design self-oscillating cilia [1]. In these systems, the chemical energy from the Belousov–Zhabotinsky (BZ) reaction is transduced into the mechanical motion of the artificial cilia [1]. We also integrated thermo-responsive gels, chemically reactive fibers, and reagent-laden fluids to devise oscillating cilia from non-oscillating components. In other examples, we combined the behavior of thermo-responsive gels and photo-responsive fibers so that both light and heat enabled the cilia-like motion of the embedded fibers. Notably, light provides a particular useful stimulus in these systems since it can be applied remotely and non-invasively; moreover, it can be readily turned on and off to control the dynamics of the cilial layers.

In order to model the behavior of the responsive gels and fibers, and the inter-conversion of the chemical and mechanical energy in the above systems, we utilized our gel Lattice Spring Model (gLSM) [2–4]. This modified gLSM approach for the gel-fiber composites [5, 6] combines finite-element and finite-difference approaches to describe the elastodynamics of the gel matrix, the bending and stretching of the anchored fibers, and the response of the system to external stimuli. Using this model, we determined how the presence of chemical reactants in the solution and variations in the external stimuli can be harnessed to regulate the global performance of the system. Moreover, this approach allows us to describe phenomena occurring on the micron to millimeter length scales.

In studies involving the gLSM, we neglected hydrodynamic effects because the movement of the gel or cilial fibers was relatively slow, and thus the fluid motion had a negligible effect on the behavior of the system. An imposed fluid flow can, however, play a vital role in regulating the dynamics and performance of artificial cilia. Consequently, we formulated a new approach to examine the effect of an applied shear flow on the ability of cilial fibers to extract particles from the solution [7]. In particular, we coupled the lattice Boltzmann model (LBM) for binary fluids with a stochastic differential equation for the motion of the nanoparticles and introduced rod-like fibers anchored to a substrate. We also incorporated the wetting interactions between the fibers and the fluids, as well as the adhesive interactions between the fibers and the particles in the solution. In this model, we assumed that the rhythmic motion of the fibers is driven by volumetric changes of the underlying gel or an alternating electromagnetic field. To simplify the model, however, we did not explicitly include the presence of the gel or field. With this approach, we can probe features on the nanometer to micron scales.

Finally, we examined a system that combined the thermo-responsive behavior of both the gel and the fibers, as well as the effects of an imposed shear flow [8]. To capture the interactions among all these components, we used the particle-based dissipative particle dynamics (DPD) simulations, which are appropriate for describing behavior occurring on length scales up to hundreds of nanometers. With the DPD, we simulated a scenario where the temperature-induced changes in both the gel and the fibers enabled the fibers to controllably "catch and release" particles in solution.

Below, we briefly describe our computational approaches and discuss the results obtained from our simulations. Overall, our findings provide guidelines for tailoring the features of the components to optimize the utility of microfluidic devices that harness hair-like structures to perform such functions as purifying contaminated solutions or targeting cells for biological assays. On a more general level, our results reveal how to combine different stimuli-responsive components to create synthetic material systems that can be controllably and repeatedly actuated to display biomimetic cilial motion and behavior.

10.2 Designing Chemo-responsive Cilial Systems

In the biological systems, active biochemical machinery drives the motion of the cilia. In our design of synthetic systems, we utilized responsive gels as the "muscle" that drives the artificial cilia. Specifically, we designed gel-based systems that transduced the energy from internalized reactions or external stimuli into mechanical motion. To carry out these studies, we utilized the gLSM [2–4], which integrates a finite-element approach to solve the equations for the gel elastodynamics and a finite-difference approximation to solve the reaction–diffusion equations at each lattice site (node) in the gel.

We originally developed the gLSM to simulate the dynamics of self-oscillating gels undergoing BZ chemical reactions [9, 10]. The BZ reaction involves a periodic reduction and oxidation of the metal-ion catalysts that are anchored to the chains in the gel. This redox reaction in turn modifies the hydrating effect of the solvent on the polymer network, which expands when the catalyst is in the oxidized state (Ru^{3+} for the cases we considered) and collapses when the catalyst is in the reduced state (i.e., Ru^{2+}). Hence, the oscillating chemical reaction gives rise to the rhythmic pulsations of the gel. With multiple, long BZ gels attached to a surface, these pulsations broadly mimic the self-sustained motion of cilia.

A number of our predictions on the behavior of the BZ gels and other chemo-responsive networks were experimentally verified [4, 11–14], making the gLSM an effective approach for simulating the dynamic behavior of chemically reactive gels. As discussed below, to further tailor our design of gel-based cilial layers, we modified the gLSM to encompass elastic, embedded fibers that extend through the surface of the gel.

10.2.1 Methodology: Gel Lattice Spring Model

Within the framework of the gLSM, we determine the velocity $\mathbf{v}^{(p)}$ of each node in the polymer network. We assume that the dynamics of the system lies in the overdamped (relaxational) regime [5], and hence the velocity is proportional to the force, F, acting on each node. The force is obtained from a spatial derivative of the total energy of the system, U. Hence, to capture the dynamic behavior of the network, we begin by first determining an expression for U.

The total energy of the chemo-responsive gels, U, is taken as the sum of the energy of the polymer-solvent interaction, U_{FH}, and the elastic energy associated with the deformation of the gel, U_{el}. The first term is written in the following Flory–Huggins form [15]:

$$U_{FH} = \sqrt{I_3}[(1-\phi)\ln(1-\phi) + \chi_{FH}(\phi, T)\phi(1-\phi) + f_{\text{int}}]. \quad (10.1)$$

Here, ϕ is the volume fraction of the polymer and $(1 - \phi)$ is the volume fraction of solvent. The first two terms in (10.1) describe the mixing energy of the gel, and the term f_{int} accounts for the interactions between the polymer network and any chemical species that are anchored to this network [2] (as discussed further below). In (10.1), $\chi_{FH}(\phi, T)$ is the Flory–Huggins polymer–solvent interaction parameter, which depends on ϕ, and temperature, T, for the thermo-responsive gels [16]. Additionally, $I_3 = \det\hat{\mathbf{B}}$ is an invariant of the left Cauchy–Green (Finger) strain tensor $\hat{\mathbf{B}}$. The prefactor $\sqrt{I_3}$ relates the change in volume of the deformed gel relative to the undeformed sample [17].

The elastic energy contribution, U_{el}, describes the rubber elasticity of the crosslinked network [15, 18], and is proportional to the crosslink density, c_0:

$$U_{el} = \frac{c_0 v_0}{2}(I_1 - 3 - \ln I_3^{1/2}). \quad (10.2)$$

Here, v_0 is the volume of a monomeric unit and $I_1 = \operatorname{tr}\hat{\mathbf{B}}$ [17].

From the above equations for U, we can derive the constitutive equation that relates the stress to the strain in the chemo-responsive gels [2]. In particular, the dimensionless stress tensor in a deformed material is given as:

$$\hat{\sigma} = -P(\phi, T)\hat{\mathsf{I}} + c_0 v_0 \frac{\phi}{\phi_0}\hat{\mathsf{B}}. \quad (10.3)$$

Here, $\hat{\boldsymbol{\sigma}}$ is measured in units of $v_0^{-1}kT$, $\hat{\mathbf{I}}$ is the unit tensor, and ϕ_0 is the volume fraction of the polymers in the undeformed state. The isotropic pressure, $P(\phi, T)$, in (10.3) is defined as

$$P(\phi, c_{\text{int}}, T) = -(\phi + \ln(1-\phi) + \chi(\phi, T)\phi^2) + c_0 v_0 \phi (2\phi_0)^{-1} + \pi_{\text{int}}. \quad (10.4)$$

The parameter $\chi(\phi)$ is related to the Flory–Huggins interaction parameter χ_{FH} in (10.1) as $\chi(\phi, T) = \chi_{FH}(\phi, T) - (1-\phi)\partial\chi_{FH}(\phi, T)/\partial\phi$. The term π_{int} is the contribution to the isotropic pressure from the last term in (10.1) and is given further below.

In the ensuing studies, we focus on poly(N-isopropylacrylamide) (PNIPAAm) gels, which display a lower critical solution temperature (LCST); for such gels,

$$\chi(\phi, T) = \chi_0(T) + \chi_1\phi, \quad (10.5)$$

where $\chi_0(T) = [\delta h - T\delta s]/kT$, and δh and δs are the respective changes in the enthalpy and entropy per monomeric unit of the gel [16].

We relate the above thermodynamic properties to the dynamic behavior of the chemo-responsive gel through the framework provided by the two-fluid model [18–20]. In the latter model, the system is assumed to be incompressible and the total velocity of the polymer–solvent system is set to zero, so that only the polymer–solvent inter-diffusion contributes to the gel dynamics [2, 21–23]. The volume fraction of polymer, ϕ, obeys the following continuity equation:

$$\partial\phi/\partial t = -\nabla \cdot (\phi \mathbf{v}^{(p)}). \quad (10.6)$$

As noted above, the dynamics of the polymer network is assumed to be purely relaxational [19], so that the forces acting on the deformed gel are balanced by the frictional drag due to the motion of the solvent. The equation that expresses this balance in forces allows us to derive the velocity of the deformed polymer network, $\mathbf{v}^{(p)}$, in terms of the strain tensor $\hat{\mathbf{B}}$. The complete derivation of this velocity is given in [2] where we also describe how this equation is solved numerically *via* the gLSM approach.

To model the artificial cilia, we extended the gLSM to account for the presence of elastic fibers that are embedded within the gel matrix. We assume that the fibers are grafted to the gel, and hence, we take the nodes comprising the fibers to be common with the gel nodes. Thus, the fibers can only move together with the gel [5].

The fiber-free energy is defined as [5]

$$U_{fil} = \sum \frac{k_b}{2}(r_{ij} - \delta)^2 + \sum \frac{k_a}{2}(\cos(\theta_{ijk}) - \cos(\theta_0))^2. \quad (10.7)$$

The first term in (10.7) represents the elastic energy of the harmonic bond between the two neighboring nodes (i, j), $r_{ij}(t)$ is the bond length between these nodes, δ is the equilibrium bond length, k_b is the spring constant, and the summation is taken over all the bonds within the fiber. The second term in (10.7) represents the angle potential [24]; here, θ_{ijk} is the angle between the two neighboring bonds sharing a common node, θ_0 is the equilibrium value of this angle set at $\theta_0 = \pi$, and the constant k_a gives the stiffness of the fiber; the summation is taken over all the pairs of bonds sharing a common fiber node. Further details of the model are given in [5] and [6].

In the following section, we briefly review our studies on chemo-responsive systems where chemical reactions within the gels or between the cilial fibers and surrounding solutions provide the energy to drive the concerted motion of the system. In the subsequent section, we describe how the introduction of chromophores into a PNIPAAm-based composite allows the cilial layer to be regulated by both light and heat.

10.2.2 Self-oscillating Cilia

Polymer gels undergoing the oscillating BZ reaction are a prime example of a biomimetic material; these BZ gels can autonomously transduce the energy from the chemical reaction within the network into periodic mechanical motion [9, 10]. Taking advantage of this distinctive behavior, Yoshida *et al.* [25] fabricated arrays of cylindrical BZ gels to form artificial cilia that display self-sustained rhythmic pulsations. In previous studies, we designed artificial cilia by anchoring one end of rectangular, *mm*-sized BZ gels to a flat substrate [1]. We found that the gels display a form of chemotaxis as the BZ cilia bend toward each other and self-aggregate [1].

Figure 10.1a shows the initial configuration of such BZ cilia, where five vertical filaments are grafted to the solid surface. The initial size of the non-deformed gels corresponds to approximately 0.2 mm $\times$ 0.2 mm $\times$ 1.3 mm, and the remaining model parameters are given in [1]. As the reaction proceeds, the activator of the BZ reaction, u (i.e., $HBrO_2$), is produced inside all the gels and diffuses into the surrounding solution. The central BZ cilium is bounded on both sides by neighbors that produce u, and hence, the highest concentration of u lies around and above this center filament (Figure 10.1b). At late times (Figure 10.1c) the anchored filaments all tilt toward the highest concentration of u in the central region. Hence, u effectively acts as a chemo-attractant and drives the anchored gels to associate. We showed [1] how the distribution of $u$in the system can be controlled by light since light of a certain

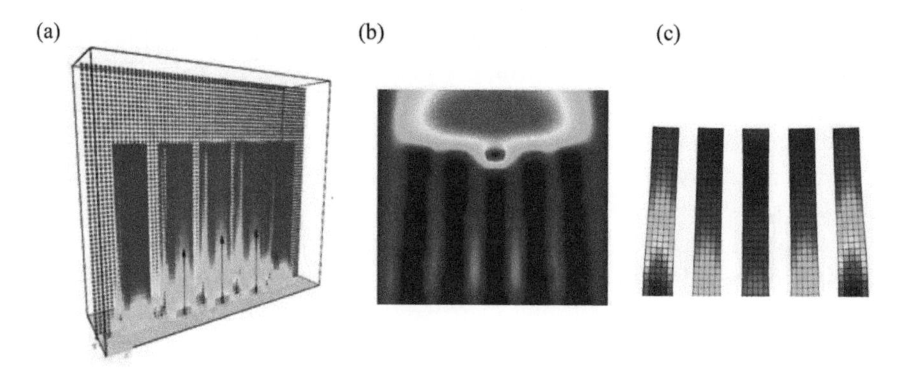

Figure 10.1 Chemically mediated communication in the system of five oscillating cilia, each cilium is made of a self-oscillating BZ gel. The size of the simulation box is $69 \times 20 \times 63$. (a) Early time snapshot system. (b) Distribution of the activator, u, in the simulation box, with the red color corresponding to the maximum concentration of u. (c) Late-time morphology of the five-cilia system. Simulation details are given in [1]. Reproduced from Reference [1] with permission from the Royal Society of Chemistry.

wavelength suppresses the BZ oscillations. Hence, non-uniform illumination with blue light can be used to modulate the self-aggregation of the cilia and, thus, tailor the dynamic behavior of the layer.

Coupled to the experimental studies [5], we also modeled a distinctly different self-oscillatory cilial array, where none of the individual components alone exhibit oscillations, but the collective interactions among these components give rise to an oscillatory regime. In the experiments [5], a bilayer of fluid was driven to flow over a PNIPAAm hydrogel. Relatively rigid microscopic fibers were embedded in and extended above this hydrogel layer. The tips of the fibers were covered with the catalysts; standing in the vertical position, these tips reached into the upper layer of the fluid, which contained reagents that underwent an exothermic reaction with the catalysts. The heat from this exothermic reaction caused the underlying LCST gel to shrink and thereby bend the fibers away from the reagent-laden upper fluid. Since the catalyst could no longer react with the reagents, the heat in the system was dissipated and the system began to cool down. This decrease in temperature consequently caused the gel to expand so that the fibers once again reached into the upper fluid. In this manner, the fibers continued to oscillate between the vertical and bent configurations. The entire system acted as a homeostatic device, keeping the temperature fixed within a relatively narrow range.

To model this complex, multi-component system, we augmented the gLSM approach to describe the interaction of the embedded fibers with the upper layer of reagents (see Figure 10.2a). The characteristic height of the gel layer in Figure 10.2 is $H = 18.6\ \mu$m, the position of the interface between the reactive and aqueous layers lies at $h = 25.14\ \mu$m, and the height of the cilia (in the unbent state) is equal to 25.62 μm; these length scales are similar to the experimental values [5]. The results from our gLSM simulations in Figure 10.2 show the temporal evolution of the z coordinate of the tips of the fibers, $z_{\mathrm{tip}}(t)$ (blue curve, right axis), and temperature, $T(t)$ (red curve, left axis). The green line marks the location of the interface between the two fluids forming the bilayer. With the increase in temperature, the LCST gel shrinks, causing the fibers to bend. At some moment in time, the tips lie below the bilayer interface, and consequently, the exothermic reaction is switched "off." After the beginning of the cooling process, the gel layer still continues to shrink for a period of time before starting to re-swell. This additional shrinking is due to the slow response of the gel to the change in temperature. Hence, the gel layer does not reach its equilibrium degree of swelling during the relatively fast heating process. Consequently, the tips' positions during the cooling cycle, $z_{tip}^{cool}(T^*)$, lie lower than the tips' positions during the heating cycle, and $z_{tip}^{heat}(T^*)$, at the same temperature, $T^*(z_{tip}^{cool}(T^*) < z_{tip}^{heat}(T^*))$ (see Figure 10.2b). The same trends were observed in the corresponding experiments [5].

This artificial ciliary system provides unique homeostatic functionality that maintains a stable temperature within the system. Hence, the system can be used to form autonomous, self-sustained thermostats, which can help regulate energy usage in various material systems. Importantly, our model provided useful predictions on controlling the performance of the device. In particular, the simulation results revealed that the homeostatic temperature and oscillation amplitude could be controlled by varying the height of the liquid–liquid interface, the geometry and mechanical properties of the fibers, and heating rate. These predicted trends were all confirmed by further experiments [5].

10.3 Photo-responsive Gel-Fiber Systems

In the above example, the cyclic conversion of chemical energy from the reaction into the mechanical motion powered the motion of the cilial layer. Using the gLSM, we also designed a system where the inter-conversion of photo-chemo-mechanical energy enabled the motion of the cilia-like fibers. These fibers again protrude from a layer of PNIPAAm gel; here, however, the fibers

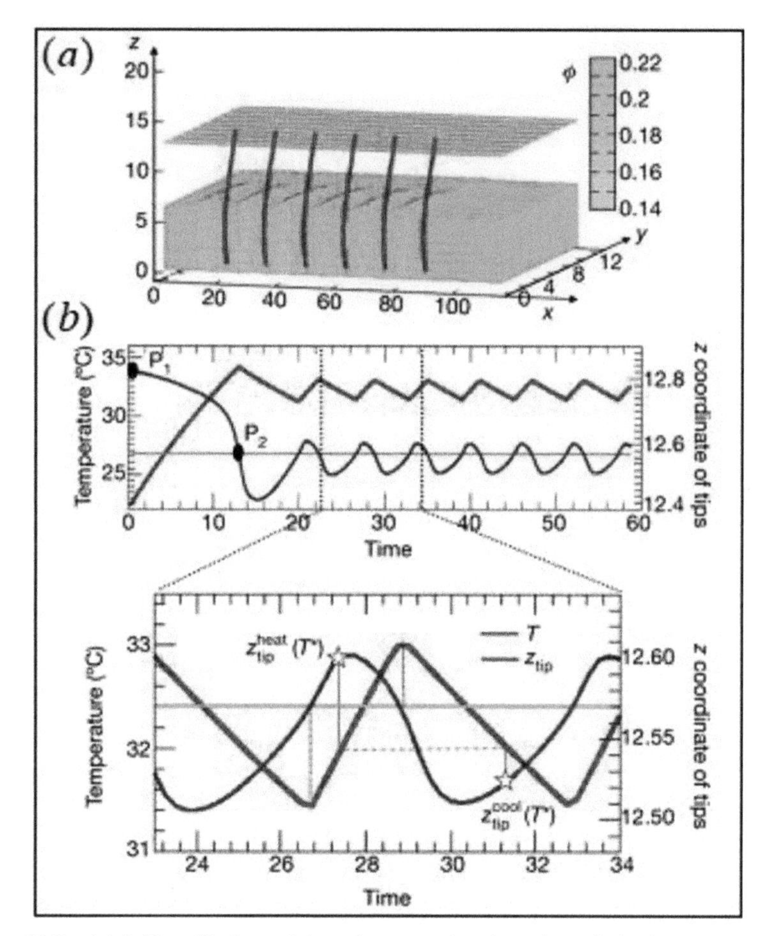

Figure 10.2 (a) Self-oscillations of the gels composite; the gel matrix is shown according to the color bar on the right that gives the volume fraction of the polymer, φ, the elastic fibers with catalyst-covered tips are shown in red, and the bilayer liquid interface is represented by the red plane. (b) Time evolution of the vertical (z) coordinate of the tips, $z_{cool}(t)$ (blue curve, right axis), and temperature, $T(t)$ (red curve, left axis). The interface position (red plane in (a)) is marked by the green line. The lower image in (b) shows an evolution during the single oscillation cycle. The positions of the tips during the cooling and heating are marked as $z_{to}^{cool}(T)$ and $z_{to}^{heat}(T)$, respectively. Reproduced from [5] with permission from Nature Publishing Group.

are functionalized with ligands that contain spirobenzopyran chromophores. The ligands on the embedded fibers are sufficiently long that chromophores extend a finite distance into the gel layer and remain anchored in the network.

In the absence of light and in acidic aqueous solutions, the spirobenzopyran chromophores are primarily in the open ring form (the protonated merocyanine form, or *McH*) and are hydrophilic [26–28]. Illumination with blue light results in the photo-conversion of these chromophores into the hydrophobic closed ring conformation (the *spiro* form, or *SP*) [26, 27]. When this light is turned off, the unstable *SP* form undergoes spontaneous conversion back to the stable, hydrophilic *McH* form:

$$McH \underset{k_D}{\overset{k_L}{\rightleftharpoons}} SP. \tag{10.8}$$

The values of k_L and k_D in (10.8) refer to the reaction rate constants for the forward and backward reactions, respectively.

The inter-conversion reaction in (10.8) is described as:

$$\frac{\partial c_{SP}}{\partial t} = k_L(I(\mathrm{r}))(1 - c_{SP}) - k_D c_{SP}. \tag{10.9}$$

In (10.9), the value of c_{SP} is normalized by the total concentration of chromophores. In the absence of light, we set c_{SP} equal to zero as an initial condition. This reaction kinetic is coupled to the gel elastodynamics within the 3D gLSM framework as detailed in [29].

Experiments on the SP-containing gels have shown that the temperature remains constant as the sample shrinks under the illumination [26]. Hence, the photo-induced shrinking of the SP-functionalized gels is due solely to the changes in the hydrophobicity of the illuminated sample and is not due to light-induced heating, which occurs with different types of chromophores [30, 31].

We considered the linear arrangement of embedded fibers depicted in Figure 10.3a. The sample corresponds to 2.3 mm $\times$ 0.3 mm $\times$ 0.5 mm in physical dimensions. The total length of the fibers is approximately 0.8 mm, with 0.3 mm extending out of the surface of the gel. (The remaining parameters are provided in [6].) Prior to the application of light or heat, the fibers are in the upright position, as shown in Figure 10.3c. We consider sufficiently thin samples so that we can neglect the attenuation of light through the thickness of the gel [29]. Initially, the sample is in the dark at $T = 20°\mathrm{C}$ with all the chromophores in the *McH* state. We note that the fibers are closer to the "front" face (at $y = 0$) of the sample than to the back and the bottom face of the composite is grafted to an underlying surface in Figures 10.3b and c.

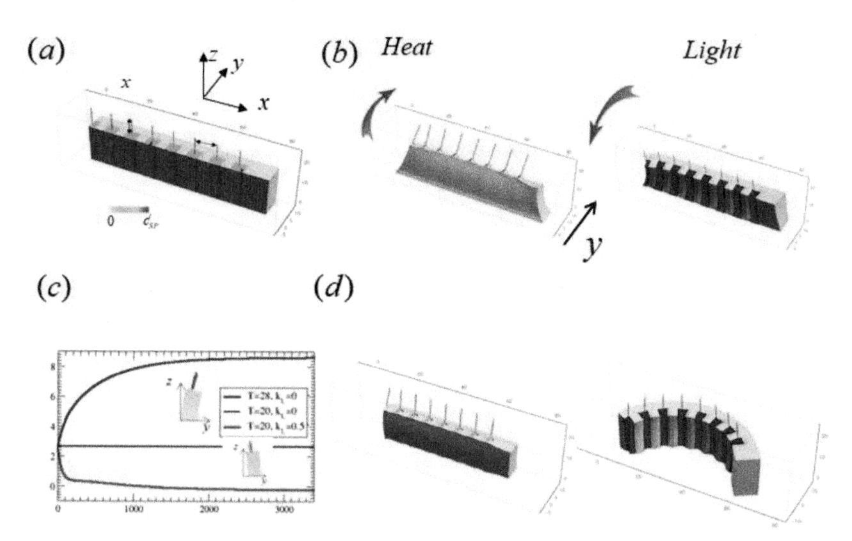

Figure 10.3 (a) Initial morphology of the gel-fiber composite with SP-functionalized fibers. d is the distance between the fibers and b is the length of the fiber extending from the surface of the gel. Green regions around the fibers mark the SP-functionalized areas. (b) Late-time morphology of the composite with the bottom surface grafted to the substrate under heat ($T = 28°$C) (left image) or under the illumination (right image). (c) Evolution of tip position in heat and light (blue and red curves as marked in the legend). Black curve corresponds to the initial tips position. (d and e) Late-time morphology of gel-fiber composite freely suspended in the solvent under heat (left image) or under the illumination (right image). Adapted from Reference [6] with permission from the Royal Society of Chemistry.

To understand the behavior of the composite in the presence of light and heat, it is useful to first recall the performance of a comparable-sized gel that does not encompass the fibers, but contains a uniform distribution of spirobenzopyran chromophores. In the latter case, the gel will undergo a spatially uniform collapse with either the application of light or heat. In the composite, however, the chromophores are not uniformly distributed throughout the gel, but are localized around the fibers. Hence, when the sample at $T = 20°$C is illuminated, the shrinking is confined only to specific regions around the elastic fibers. This localized shrinking produces the pattern observed in the image on the right in Figure 10.3b, where the fiber bends toward the negative y-direction [6].

In contrast, when the initial sample in Figure 10.3a is heated to $T = 28°$C in the absence of light, the fibers bend in the positive y-direction (left image in Figure 10.3b). In other words, the fibers are deflected in opposite directions when the composite is exposed to the different stimuli.

The common feature in both the illuminated and heated samples is that the fibers bend toward the more collapsed regions. The more collapsed region, however, is different in the case of exposure to heat and exposure to light. With heat (and in the absence of light), the back of the heated sample is more collapsed because the relatively stiff fibers inhibit the shrinking near the front. (Recall that the fibers are closer to the front of the gel.) This non-uniform response results in the entire sample bending in the positive y-direction (right image in Figure 10.3b). On the other hand, with light (and no heat), the collapse of the gel is localized near the fibers at the front of the sample, and hence, the fibers bend in the negative y-direction.

Figure 10.3c shows the time evolution of the y-coordinate of the tip of the central fiber. The blue and red curves correspond to the respective heated (with $T = 28°C$) and illuminated samples in Figure 10.3, while the black line indicates the y-coordinate of the tip of the same fiber for the sample in the dark at $T = 20°C$. The images in Figure 10.3 indicate that light and heat can be used as orthogonal stimuli to regulate the shape of the composite.

Samples that are not grafted to the surface can also be regulated by the application of heat or light [6]. When the sample is free in solution, the bottom face is no longer constrained, and thus this area also responds to the stimulus. Hence, when the composite is heated, it undergoes a uniform shrinkage. Given an initial configuration similar to that in Figure 10.3a, the heated sample resembles a compressed accordion, where the fibers are brought into closer contact by the lateral collapse of the gel (see Figure 10.3d). If the sample is illuminated, the collapse of the SP-containing bands drives the entire composite to bend into a semi-circle. With the fibers extending from the surface of the gel, the system now resembles a bent, hairy caterpillar (see Figure 10.3e).

10.4 Capturing Hydrodynamic Interactions in Complex Fluids

In the above examples, the behavior of the cilial layer was controlled by the transduction of energy within the gel matrix. In the latter studies, we neglected the hydrodynamic interactions within the system and the role that imposed flows can play in regulating the dynamics of the cilial layer. In the following section, we summarize our recent work on modeling the effect of an imposed flow on rod-like cilia, which controllably catch, transport, and release specific target molecules within the surrounding solution, and, thus, could be harnessed for effective separation processes within microfluidic devices [7]. To optimize this functionality, we considered the cilia to be immersed in two immiscible flows, which form two distinct streams (see Figure 10.4). We assumed that the rods underwent externally driven oscillations, enabling these rods to transport the targeted species from the upper stream and release them to the lower stream, where they could ultimately be extracted from the device. The rods' oscillations could be induced by the expansion and contraction of an underlying gel (as discussed above) or the application of an oscillatory magnetic field to magneto-responsive rods. We did not, however, explicitly model the presence of the gel or the electromagnetic field in these studies [7].

The target species in this system are modeled as adhesive nanoparticles that are initially dispersed in the upper solution. Only the rod sites that extend into the upper stream are "activated," and thus, can "catch" (bind) the adhesive nanoparticles. The binding interaction involves the formation of bonds between the activated rod sites and the particles. The oscillating rods are driven to move back and forth between the two fluid streams. Thus, when these sites are brought into the lower fluid, the bonds are broken and the respective rod sites become deactivated, rendering them non-adhesive to both types of nanoparticles. To determine the efficiency of this device, we determined the fraction of adhesive particles that the rods can "catch" relative to the total particles in the system.

10.4.1 Methodology

The ensuing simulations were carried out in 2D, and thus the cilia were modeled as rods (see Figure 10.4) [7]. The model does, however, capture fundamental features of 3D experimental systems where, for example, the cilia are fins that are broad in the third dimension (i.e., into the page). This array of rods is anchored to the floor of the microchannel and an immiscible 50:50 AB binary mixture is driven to flow through the microchannel by an imposed gradient pressure. The incompressible AB binary mixture is characterized by the continuous order parameter, ϕ, defined as the difference between the local mass density of the A and B components: $\phi = (\rho_A - \rho_B)/\rho$, where $\rho = \rho_A + \rho_B$ is the total mass density of the fluid [32]. The values $\phi = 1$ and -1 correspond, respectively, to the A-rich (lower red) and B-rich (upper blue) phase (see Figure 10.4a).

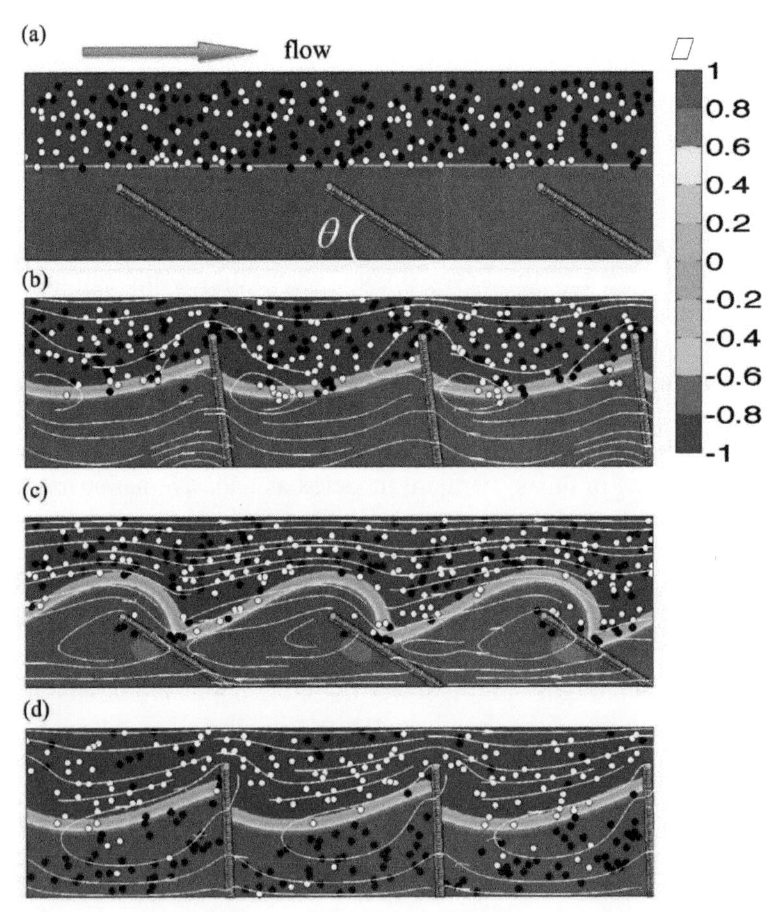

Figure 10.4 Snapshots of the system, illustrating mechanism of oscillating fin "catching and releasing" particles within a binary blend. (a) The initial morphology of the system with two types of particles, adhesive (in black) and non-adhesive (in white), randomly distributed in the upper stream. The blend is represented by the value of order parameter $\phi(\mathbf{r})$ as illustrated by the color bar. The oscillating angle θ varies from $\pi/6$ to $\pi/2$. (b) At $t = 3 \times 10^4$, fins catch adhesive particles in the upper stream, where the fluid streamlines are shown in white. (c) At $t = 3.2 \times 10^5$, attached adhesive particles are released from fins into lower stream. (d) Late-time morphology of the steam at $t = 5 \times 10^6$. Reproduced from Reference [7] with permission from the Royal Society of Chemistry.

The oscillating rods reach into the upper fluid (B-rich phase) when they are upright and are entirely immersed within the lower stream (A-rich phase) when they are tilted. At the onset of the simulations, we introduced a binary mixture of mobile nanoparticles into the upper fluid: the black nanoparticles can adhere to the rods, while the white particles are non-adhesive (Figure 10.4a). The portion of the rods that extend into the upper B stream can bind the neighboring adhesive particles.

The free energy functional that describes the nanoparticle-filled AB binary mixture can be written as [32]:

$$F = \int \left[\psi_b(\phi, \rho) + \frac{\kappa}{2}|\nabla\phi|^2\right] \mathbf{dr} + \int \psi_s(\phi_s)\mathbf{ds} \tag{10.10}$$

where $\psi_b(\phi, \rho)$ is the Landau–Ginzburg free energy for the binary fluid and is given by:

$$\psi_b(\phi, \rho) = \frac{1}{3}\rho \ln \rho - \frac{a}{2}\phi^2 + \frac{b}{4}\phi^4. \tag{10.11}$$

The second term in (10.10) characterizes the energy penalty for spatial variations of the order parameter and is thus related to the interfacial tension $\sigma = \sqrt{8\kappa a^3/9b^2}$. When the parameters a, b, and κ are positive constants, the binary mixture spontaneously phase-separates into A-rich and B-rich domains with the interfacial width equal to $\xi = 5\sqrt{\kappa/2a}$. The equilibrium value of the order parameters is $\phi = \sqrt{a/b}$ and $\phi = -\sqrt{a/b}$ for A-rich and B-rich phases, respectively.

The second integral in (10.10) represents the wetting interactions at the fluid–solid boundaries that depend on the distribution of the order parameters on the interface ϕ_s: $\psi_s(\phi_s) = h\phi_s$ [33], where h is a tunable parameter that characterizes the strength of the wetting interaction. The value of h is related to the static contact angle θ_{st} [34, 35]:

$$h = \sqrt{2\kappa b}\, \mathrm{sgn}(\pi/2 - \theta_{st})\sqrt{\cos(\alpha/3)(1 - \cos(\alpha/3))}, \tag{10.12}$$

where $\mathrm{sgn}(x)$ is the sign function and $\alpha = \arccos(\sin^2\theta_{st})$. In our system, we specify the wetting interactions between the following: 1) the binary fluids and the top and bottom walls of the microchannel and 2) the binary fluids and the rods. By minimizing the free energy functional (10.12) with respect to the order parameter at the boundary between the rods and the binary mixture, we obtain the following boundary condition [36]:

$$\mathbf{n} \cdot \nabla\phi = h/\kappa, \tag{10.13}$$

where $\mathbf{n}$ is the unit vector normal to the interface.

The temporal evolution of the order parameter is governed by the following convection–diffusion equation:

$$\frac{\partial\phi}{\partial t} + \nabla \cdot (\phi\mathbf{u}) = M\nabla^2\mu \tag{10.14}$$

where M is the mobility of the order parameter, μ is the chemical potential related to the free energy functional: $\mu = \delta F/\delta\phi$, and $\mathbf{u}(\mathbf{r}, t)$ is the fluid velocity. The dynamics of nanoparticle-filled fluid is described by the Navier–Stokes equation:

$$\frac{\partial}{\partial t}(\rho\mathbf{u}) + \nabla \cdot (\rho\mathbf{u}\mathbf{u}) = -\nabla \cdot \mathbf{P} + \eta\nabla^2\mathbf{u} + \mathbf{G}^{drag}, \tag{10.15}$$

where **P** is the pressure tensor and η is the viscosity of the fluid, and the last term, $\mathbf{G}^{drag}$, arises from viscous drag acting on the nanoparticles [33]:

$$\mathbf{G}^{drag}(\mathbf{r}) = -\sum_i \delta(\mathbf{r} - \mathbf{r}_i)\mathbf{F}^{drag,i} = -\zeta[\dot{\mathbf{r}}_i(t) - \mathbf{u}(\mathbf{r}_i, t)] \tag{10.16}$$

where $\mathbf{F}^{drag,i} = -\zeta[\dot{\mathbf{r}}_i(t) - \mathbf{u}(\mathbf{r}_i, t)]$ is the frictional drag force on the ith nanoparticle with the friction coefficient being $\zeta = 6\pi\eta R_p$. The radius of the nanoparticles is sufficiently small that we model these species as tracer particles. The dynamic behavior of these nanoparticles is governed by the following stochastic differential equation:

$$d\mathbf{r}_i(t) = \mathbf{u}(\mathbf{r}_i, t)dt + \sqrt{2D_p}d\mathbf{W}_i(t) + \frac{dt}{\zeta}[\mathbf{F}_i^e(t) + \mathbf{F}_i^a(t)]. \tag{10.17}$$

The first term on the right-hand side accounts for the drift velocity due to the fluid motion. The second term represents the random force acting on the ith particles and satisfies the fluctuation–dissipation relation: $\langle \mathbf{W}_i(t) \cdot \mathbf{W}_j(t') \rangle = 2k_b T D_p \delta_{ij}\delta(t - t')$. The term $\mathbf{F}_i^e(t)$ in (10.17) represents the nanoparticle–nanoparticle and nanoparticle–rod excluded volume interactions and is derived from the repulsive part of Morse potential [37]:

$$\psi_m(r) = \varepsilon(1 - \exp[-\lambda(r - r_c)])^2, \tag{10.18}$$

where the values of ε and λ derode the strength and range of the potential, and r_c is the relevant equilibrium separation. The parameter $\psi_m(r)$ is repulsive if $r < r_c$ and attractive if $r > r_c$; to prevent the overlap between the particles, and between the particles and the rods, the potential for the excluded volume interaction is applied when $r < r_c$. The force $\mathbf{F}_i^a(t)$ in (10.17) accounts for the attractive interaction between the rods and adhesive particles in the upper stream.

Driven by an externally applied force, the rods oscillate between $\theta_{\min}$ and $\theta_{\max}$, the respective minimum and maximum angles formed between the rods and floor of the microchannel. Given that the period of oscillations is T, the temporal variation in the angle between the rods and floor is given by:

$$\theta(t) = \frac{1}{2}(\theta_{\max} + \theta_{\min}) - \frac{1}{2}(\theta_{\max} - \theta_{\min}) \cos \frac{2\pi t}{T}. \tag{10.19}$$

The corresponding angular velocity of the rods is given by $\omega(t) = \frac{\pi}{T}(\theta_{\max} - \theta_{\min}) \sin \frac{2\pi t}{T}$; the value of ω is zero when $\theta(t)$ reaches $\theta_{\min}$ or $\theta_{\max}$.

To model the selective attraction in the system, we introduce a bond-like interaction between the adhesive nanoparticles and the sites on the rods located in the upper stream. In particular, an attractive Morse potential with the same form as (10.18) and given by $\psi_a = \varepsilon(1 - \exp[-\lambda(|\mathbf{r}_p - \mathbf{r}_f| - r_c)])^2$ is applied when $r_c \leq |\mathbf{r}_p - \mathbf{r}_f| < r_b$, where $\mathbf{r}_p$ and $\mathbf{r}_f$ are the respective positions of the centers of the adhesive particles and the rod sites [38]. If the separation between adhesive nanoparticles and the rods is greater than the bond length, i.e., when $|\mathbf{r}_p - \mathbf{r}_f| \geq r_b$, the above potential is no longer applied; when $|\mathbf{r}_p - \mathbf{r}_f| < r_c$, the excluded volume interaction dominates. An adhesive site on a rod can form only a single bond with the adhesive nanoparticle, and hence the maximum number of nanoparticles that can potentially be collected by the rods is limited to a specific value. In the simulations below, we set the parameters to the following values: $\lambda = 0.8$, $\varepsilon = 1.5$, and $r_c = 1.2$.

Equations (10.13) and (10.15) are solved by using the LBM [39, 40], which has proven to be an efficient method for modeling the complex fluid dynamics in systems that encompass multiple length scales. We combine the bounce-back scheme and interpolations of the order parameter on the moving boundary to implement the boundary conditions on the rods [41–43].

The channel size is 40×150 in dimensionless units, with periodic boundary conditions applied on the horizontal direction and bounce-back conditions on the top and bottom of the channel. The static contact angle between the respective fluid and the walls of the channel is set to $\theta_{st}^{w} = \pi/3$. The respective values for the characteristic unit of length and time in our simulations are: $L_0 \approx 0.8$ μm and $T_0 \approx 10^{-7}$s. Consequently, the nanoparticle radius is approximately 77 nm. The length of rods is $L_{fin} = 24$ μm, the height of the channel is $W = 32$ μm, and the oscillation frequency is $f = 66$ Hz. The fluid velocity measured directly from the simulations does not exceed 3×10^{-3}, which corresponds to 2.4×10^{-2} m/s.

10.4.2 Utilizing Oscillating Rods to "Catch and Release" Targeted Nanoparticles

At the outset of these simulations, the adhesive (black) and non-adhesive (white) nanoparticles are randomly distributed solely within the upper fluid stream (Figure 10.4a). The imposed pressure gradient drives the fluid to flow from left to right within the channel. We first considered a total of $N = 300$ nanoparticles ($N_a = 150$ adhesive and $N_n = 150$ non-adhesive species). The three equally spaced rods on the bottom of the channel are initially fully immersed in the lower stream with $\theta = \theta_{\min} = \pi/6$ (Figure 10.4a). The maximal angle formed between the rods and bottom wall is $\theta_{\max} = \pi/2$. The preferential wetting interactions between the rods and the fluids are controlled by the static contact angle $\theta_{fin} = \pi/2.5$ through (10.12) and (10.13).

The transport of nanoparticles within this system can occur through two predominate mechanisms. Mechanism 1 involves the selective capture of the targeted adhesive nanoparticles through their attraction to the rod in the upper stream, their transport into the lower stream, and the subsequent breakage of the bonds. Mechanism 2 involves the motion of nanoparticles along the streamlines generated by the combination of the external pressure gradient and the rods' oscillatory motion; depending on the resulting flow profiles, these streamlines could bring both types of nanoparticles (adhesive and non-adhesive) into the lower fluid or pump them back into the upper fluid. The diffusion coefficient of the nanoparticles ($D_p \approx 6.4 \times 10^{-11}$ m^2/s) is sufficiently small compared to the motion of the fluids from the imposed flow and the rods' oscillations that the contribution from the diffusive motion of the particles is significantly smaller than the contributions from mechanisms 1 and 2.

Mechanism 1 depends not only on the range and strength of the adhesive interaction but also on the number density of the nanoparticles and availability of the adhesive rod sites in the upper fluid stream. As illustrated in Figure 10.4b, rod sites within the upper stream can successfully catch the adhesive particles. As the rods move downward, the bound adhesive particles follow the motion of rods and are delivered into the lower stream; thereafter, the bonds are broken and those adhesive particles are released as illustrated in Figure 10.4c.

Due to mechanism 2, however, a small number of non-adhesive particles are also delivered into the lower stream (see the white particles in Figure 10.4c). The flow close to the rods is determined mainly by the motion of the rods. Therefore, mechanism 2 is non-selective and delivers both types of nanoparticles into the lower stream.

By tuning the interplay between mechanisms 1 and 2, we can optimize conditions where the majority of the adhesive particles are brought into the lower stream and the majority of non-adhesive particles are localized in the upper stream (as shown at $t = 5 \times 10^6$ in Figure 10.4d). In particular, the interplay between these mechanisms can be modified by varying the value of the static wetting angle θ_{fin} (see Figure 10.5). By comparing the dynamics of the system with large wetting angle $\theta_{fin} = \pi/2.5$ (Figure 10.4d) and with $\theta_{fin} = \pi/5$ (Figure 10.5b), we observe significantly larger distortions of the interface for the system with the larger θ_{fin}. Furthermore, the fluid–fluid interface at the larger θ_{fin} displays less curvature. Consequently, more of the rod sites are surrounded by the upper fluid for the system with the larger θ_{fin}, and hence more sticky sites have access to adhesive particles in this case than in the scenario with the lower value of θ_{fin}.

The curves in Figure 10.5a display the temporal evolution of $C_a(t)$ and $C_n(t)$, which are the respective ratios between the number of adhesive and non-adhesive nanoparticles in the lower stream, relative to the total number of these particles, for the three different wetting angles: $\theta_{fin} = \pi/2.5$ (black), $\pi/5$ (red), and $\pi/10$ (green). The solid curves represent the average values over eight independent simulation runs and the shadings represent the standard deviations. The period of oscillations in $C_a(t)$ and $C_n(t)$ is equal to the rods' oscillation period. The value of $C_a(t)$ increases in all cases, indicating that the number of adhesive particle brought down into the lower phase during the forward stroke always exceeds the number of the particles that are able to escape from this lower phase during the recovery stroke. At late times, $C_a(t)$ approaches saturation and larger wetting angles lead to a greater value of $C_a(t)$, which indicates that the adhesive particles are delivered into the lower stream with greater efficiency at larger θ_{fin}.

The late-time images of the system for $\theta_{fin} = \pi/2.5$ (Figure 10.4d) and $\theta_{fin} = \pi/5$ (Figure 10.5b) confirm these observations. These snapshots also show that there are no non-adhesive particles within the lower stream for $\theta_{fin} = \pi/5$ at $t = 5 \times 10^6$, while a small portion of non-adhesive particles is found for $\theta_{fin} = \pi/2.5$ at the same time. This difference is reflected in the values of $C_n(t)$ for these two cases (Figure 10.5a). For larger wetting angles, $C_n(t)$ increases and reaches a saturation value around 0.1, which is 10 times higher than the value of $C_n(t)$ for $\theta_{fin} = \pi/5$. Hence, at the largest wetting angle considered here, one gains greater efficiency (larger $C_a(t)$) but loses

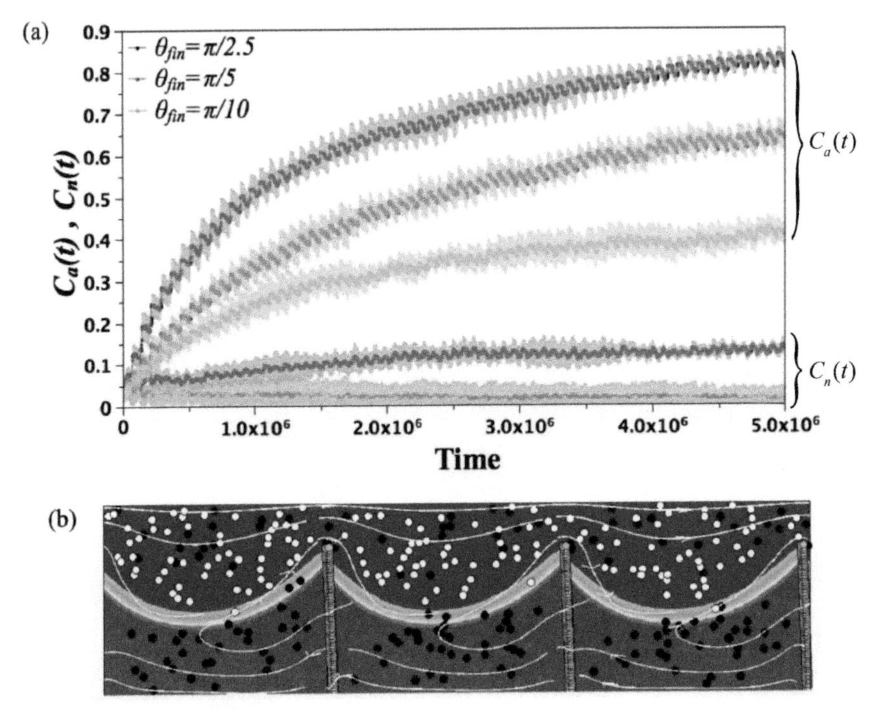

Figure 10.5 (a) Temporal evolution of $C_a(t)$and $C_n(t)$ for $\theta_{\text{fin}} = \pi/2.5$ (black), $\pi/5$ (red), and $\pi/10$ (green) with $N_a = N_n = 150$. (b) Late-time morphology of the system with $\theta_{\text{fin}} = \pi/5$ at $t = 5 \times 10^6$.

selectivity (larger $C_n(t)$) in the release of the nanoparticles into the lower stream.

To characterize the selectivity of the process, we introduce the parameter β, which is the ratio of the number of non-adhesive particles to the number of adhesive particles within the lower stream. Here, $N_a = N_n$, and hence, $\beta = C_n/C_a$. The process is non-selective when $\beta = 1$ and is the most selective when $\beta = 0$. We find that $\beta = 0.16$, 0.018, and 0.021 for $\theta_{fin} = \pi/2.5$, $\pi/5$, and $\pi/10$, respectively. These values indicate that one can obtain relatively high efficiency without sacrificing the selectivity of the delivery process at the optimal wetting angle $\theta_{fin} = \pi/5$.

In the final section below, we describe another cilial system that can be used for the catch and release of particles in flowing fluids. In these systems, we utilized the combined effects of responsive gels, deformable fibers, and an imposed flow to achieve optimal functionality.

10.5 Artificial Cilia Based on Biological Aptamers

With the different course-grained approaches, we could explicitly model the responsive gels (*via* the gLSM) or the imposed flow (*via* the LBM) that actuated the cilia. These methods allowed us to capture phenomena on the micro- and meso-scales. In recent studies, we utilized a particle-based approach, the DPD, to model the actuating gel, driven fluid, and thermo-responsive cilia in order to design a device that can controllably trap and release particles in solution in response to variations in temperature [8]. While the DPD allows us to simulate all of these components, we are now constrained to focus on a smaller length scale, namely, the DPD is applicable to lengths up to hundreds of nanometers (and time scales on the order of tens of microseconds [3]). Nonetheless, the fundamental concepts and governing principles uncovered in these studies are also applicable on the microscale.

In the systems described in this section, the hair-like, flexible fibers are end-grafted to an underlying LCST gel and the applied shear flow transports buoyant nanoparticles through the device (Figure 10.6). These fibers mimic the temperature-dependent behavior of biological aptamers, which can be driven to controllably change conformation from a folded to unfolded structure by varying the local pH of the solution [44] or the temperature of the system [45]. The biological activity of aptamers is directly related to their structure: folded aptamers can bind and trap molecules in solution, but these bonds are broken when the aptamers unfold. Importantly, the biological activity of the aptamers can be restored by altering the pH or temperature to produce the folded structure [44, 45]. Mimicking these properites, the fibers in our system form a hairpin structure at low temperatures (T) and unfold at higher T, losing their binding affinity.

The thermo-responsive behavior of the aptamers has been exploited to perform the selective catch-and-release of biological cells in microchambers. The aptamers were bound to a solid substrate whose temperature could be carefully regulated [45]. At relatively low temperatures, the aptamers folded into a hairpin-like structure and, hence, could bind the cells that were driven to flow through the device. The temperature was then increased to 48°C, which lies above the transition temperature associated with the unfolding of the aptamers. At this elevated temperature, the aptamer–cell binding was disrupted and the cells were released to the surrounding fluid.

The novelty of our system is that it exploits the synchronized, temperature-dependent behavior of both the fibers and an underlying polymer gel [8]. Namely, the expansion of the gel at low temperatures effectively

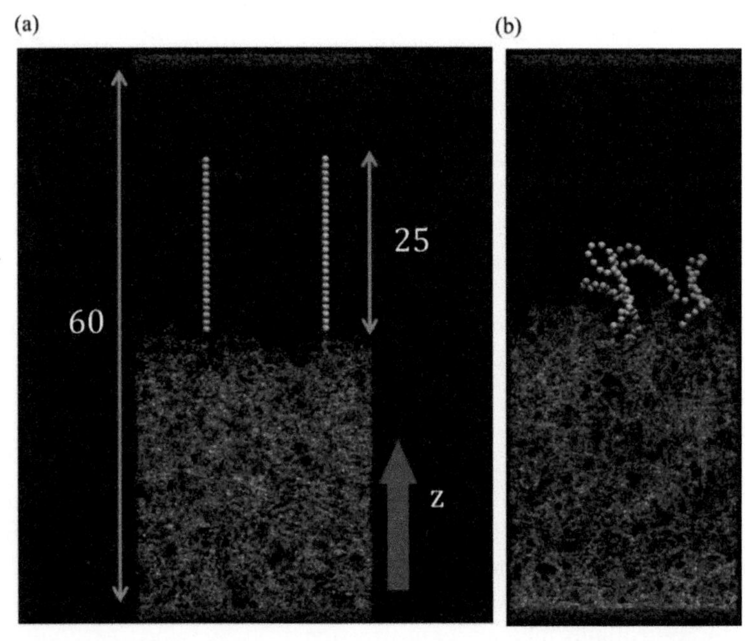

Figure 10.6 (a) Initial morphology of the system. (b) Snapshot of the system at $t = 10^6$ after equilibration at temperature equal to $T = 28°\text{C}$. Reproduced from Reference [8] with permission from the American Chemical Society.

pushes the anchored, hairpin-shaped chains into the path of the fluid-driven particles, and thereby enables the system to trap and extract these particles from the fluid stream (see Figure 10.6). In this manner, the swollen gel plays an active role in optimizing the performance of the system. As the system is heated above $T = 40°\text{C}$, the unfolding of the chain and the collapse of the supporting gel enable the applied shear flow to dislodge the particles and transport them away from the layer. The shrinking of the gel at high T also contributes to the utility of the system since it increases the size of the region affected by the flowing fluid. Since the temperature-induced conformational changes of the chains and polymer gel are reversible, the system can be used to repeatedly catch and release the particles.

To model this system, we integrated a DPD method for simulating the conformations of proteins [46] with a recently developed DPD model for thermo-responsive gels [47]. The former method [46] reproduces transitions among coil-like, globular, α-helical, and β-hairpin configurations of model peptides. The latter model [47] accurately captures the temperature-induced volume phase transition for an LCST polymer gel [16]. By integrating these two DPD schemes, we could probe the concerted response of the gel and aptamer to variations in temperature.

Below, we first briefly summarize our new DPD scheme. We then describe how the thermal response of the gel/fiber composite can be utilized for effective catch-and-release applications. As discussed below, the findings from these studies can be used to design microfluidic systems that trap and release biological cells, as well as nanofluidic devices that catch and release nanoparticles.

10.5.1 Methodology

To model the dynamic behavior of this multi-component system, we use DPD [48–50], which captures the time evolution of a many-body system via the numerical integration of Newton's equation of motion, $m\, d\mathbf{v}_i/dt = \mathbf{f}_i$, where the mass m of a bead is set equal to one. The total force acting on each bead is the sum of three pair-wise additive forces: $\mathbf{f}_i(t) = \sum(\mathbf{F}_{ij}^{\mathrm{C}} + \mathbf{F}_{ij}^{\mathrm{D}} + \mathbf{F}_{ij}^{\mathrm{R}})$. The sum runs over all beads j within a certain cutoff radius r_{c}. The conservative force is a soft, repulsive force given by $\mathbf{F}_{ij}^{\mathrm{C}} = a_{ij}(1 - r_{ij})\hat{\mathbf{r}}_{ij}$, where a_{ij} is the maximum repulsion between beads i and j, $r_{ij} = |\mathbf{r}_i - \mathbf{r}_j|/r_{\mathrm{c}}$, and $\hat{\mathbf{r}}_{ij} = \mathbf{r}_{ij}/|\mathbf{r}_{ij}|$. The drag force is $\mathbf{F}_{ij}^{\mathrm{D}} = -\gamma\omega_{\mathrm{D}}(r_{ij})(\hat{\mathbf{r}}_{ij} \cdot \mathbf{v}_{ij})\hat{\mathbf{r}}_{ij}$, where γ is a simulation parameter related to viscosity, ω_{D} is a weight function that goes to zero at r_{c}, and $\mathbf{v}_{ij} = \mathbf{v}_i - \mathbf{v}_j$. The random force is $\mathbf{F}_{ij}^{\mathrm{R}} = \sigma\,\omega_{\mathrm{R}}(r_{ij})\xi_{ij}\hat{\mathbf{r}}_{ij}$, where ξ_{ij} is a zero-mean Gaussian random variable of unit variance and $\sigma^2 = 2k_{\mathrm{B}}T\gamma$. We use $\omega_{\mathrm{D}}(r_{ij}) = \omega_R(r_{ij})^2 = (1 - r_{ij})^2$ for $r_{ij} < r_{\mathrm{c}}$ [48]. Each of these three forces conserves momentum locally, so a system of even a few hundred particles displays hydrodynamic behavior [51, 52]. The equations of motion are integrated in time via the standard velocity-Verlet algorithm. We take r_{c} as the characteristic length scale and $k_{\mathrm{B}}T$ as the characteristic energy scale in our simulations; both are set equal to one in the ensuing studies. The characteristic time scale is then defined as $\tau = \sqrt{mr_{\mathrm{c}}^2/k_{\mathrm{B}}T}$. The remaining simulation parameters are $\sigma = 3$ and $\Delta t = 0.02\tau$ with a total bead number density of $\rho = 3$ [48].

With respect to capturing the temperature-dependent behavior of the gels, at the onset of the simulations, a tetra-functional network is arranged in a diamond-like lattice structure [53]. The polymer strands between the crosslinks are modeled by a sequence of $N = 30$ DPD beads connected by harmonic spring-like bonds, with an angle potential between two consecutive bonds. Taking the values for the relevant interaction parameters from experimental data for PNIPAAm [54–58], our model accurately reproduce the continuous volume phase transition occurring between 30 and 35°C, capturing the fact that PNIPAM exhibits an LCST phase transition [16, 47, 59]. Our approach was the first DPD simulation to capture temperature-induced volume phase transitions in gels.

The thermo-responsive fiber was modeled as a flexible polymer chain connected by the harmonic bonds. To capture the conformational change of the chain such that it adopts the hairpin-like structure at $T < T_{\mathrm{M}}$ and an ideal-chain structure at $T > T_{\mathrm{M}}$, we introduced two additional types of bonds. The first is a harmonic bond connecting beads separated by two bonds. Second, we introduced a temperature-dependent Morse bond connecting beads separated by three or more bonds. This overall approach for modeling aptamers is based on DPD simulations of polypeptides in solution that successfully reproduced transitions among the different configurations of model peptides.

In our simulations, a polymer bead represents 1.6 PNIPAAm monomers. The characteristic time scale is equal to $\tau_{\mathrm{DPD}} = 0.87$ns, the simulation box size is $23.9 \times 23.9 \times 58.2$ nm, the fiber equilibrium length is 27.2 nm, and the particle diameter is 0.97nm. The diffusion coefficient of the nanoparticle corresponds to $D \approx 2.7 \times 10^{-13}\mathrm{m}^2/\mathrm{s}$. The value of the applied shear rate ranges from $\dot{\gamma} = 1.9 \times 10^5/\mathrm{s}$ to $1.9 \times 10^7/\mathrm{s}$; these shear rates are of the same order of magnitude as those used in high shear rate experiments in microfluidic devices [60, 61]. For these ranges of shear, the corresponding Peclet number can be roughly estimated as $\mathrm{Pe} \in (8, 400)$, and hence the motion of the nanoparticle is dominated by advection [62].

10.5.2 Catch and Release with Aptamer-mimicking Cilia

In a device based on these principles, the gel layer would be anchored to the floor of a microfluidic chamber, with the aptamers attached to the top of this layer. We specify that the aptamer-like chains are incompatible with the gel chains so that these aptamers do not penetrate into the gel, but rather extend into the fluid. A pump would drive a solution containing the targeted particles past this aptamer-functionalized gel. The temperature of the system could be controlled externally [45]. As the system is cooled, the hairpin-shaped aptamers on the swollen gel catch and trap the particles; as the system is heated, the unfolding of these chains on the collapsed gel would lead to the release of the trapped species.

For the aptamers, the unfolding of the chain leads to a loss of biological activity, i.e., an inability to bind the targeted molecules. In our simulations, we mimic this behavior by increasing the repulsion between a fiber and a particle bead when the chain is unfolded. Hence, when the hairpin unfolds, a bound particle can readily detach from the fiber and be removed by the imposed shear flow. We exploit this behavior in our gel/fiber composite.

At the outset of the simulations, the polymer gel is equilibrated at the temperature $T = 28°C$ and four thermo-sensitive fibers are aligned along the z-direction (Figure 10.6a). Figure 10.6b shows the equilibrium configuration of the system after it was equilibrated for 2×10^6 time steps at the same temperature ($T = 28°C$); since the fibers are incompatible with the gel ($a_{ij} = 40$ in the equation for $\mathbf{F}^C_{ij}$), the hairpins extend away from the gel layer and into the solution.

A particle with a diameter of 10 dimensionless units is introduced into the simulation box. We apply a shear flow (from the left to right of the simulation box along the x-direction) with the rate $\dot{\gamma} = 1.67 \times 10^{-3}$ to propel the particle through the system. At late times (Figure 10.7a) all four hairpins are wrapped around the particle, arresting and trapping it by a large number of hairpin–particle contacts.

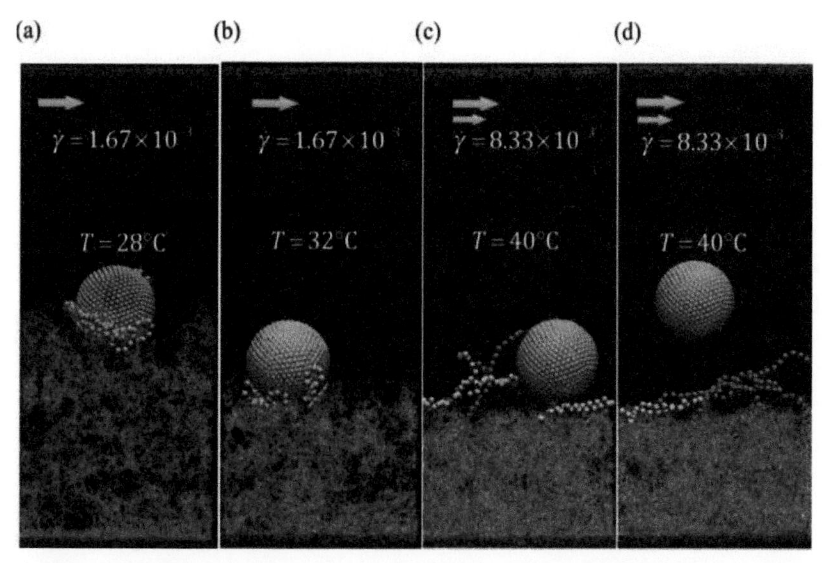

Figure 10.7 (a) Initial morphology of the system at the low temperature of $T = 28°C$. (b) Snapshot of the system at $t = 3 \times 10^6$ after the temperature increases to $T = 32°C$. The gel shrinks with the particle arrested by the hairpin. (c) The system with morphology shown in (b) is heated to $T = 40°C$ and the snapshot corresponds to $t = 3.024 \times 10^6$ under the shear rate $\dot{\gamma} = 8.33 \times 10^{-3}$. (d) Snapshot of the system at $t = 3.188 \times 10^6$.

To illustrate the merit of utilizing the thermos-responsive gel in the composite, we first increase the temperature to $T = 32°\text{C} < T_M$. At this temperature, the LCST gel shrinks, but the fibers remain in the hairpin structure. Hence, the particle remains bound to the fibers and is drawn downward toward the substrate by the collapsing gel, as illustrated in Figure 10.7b, which shows a snapshot of the system at $t = 3 \times 10^6$. In the second stage of the heating process, we further increase the temperature above the transition temperature T_M and apply a stronger shear flow with the rate of $\dot{\gamma} = 8.33 \times 10^{-3}$. Therefore, the hairpins unfold (see Figures 10.7c and d) and simultaneously lose their affinity to the particle. Due to the applied shear, the fibers are stretched along the shear direction. At late times, the particle becomes detached from the fibers and is released to the bulk solution (Figure 10.7d) [63]. The shrinking of the thermo-responsive gel plays an important role because it increases the unobstructed volume that is available for the flowing fluid, which transports the particle away from the layer.

The behavior of the system can be reset by decreasing the temperature. In particular, the gel expands and the chains refold into the hairpin structure as the temperature is decreased to $T = 28°\text{C}$. (The conformation and activity of certain aptamers can be recovered through a decrease in temperature) [45]. Hence, the temperature of the system can be cycled to promote the trapping of the particles at low temperature and their release at high temperature.

We further emphasize the unique advantages offered by matching the thermo-responsive behavior of the gels and fibers in the reversible catch-and-release of particles; we performed the studies illustrated in Figure 10.8. Here, the system's temperature is fixed at $T = 40°\text{C}$ and the gel is equilibrated in the collapsed state for 10^6 time steps. We then introduce the particle, fibers, and apply an imposed shear flow. Initially, the fibers are taken to be vertically aligned and unfolded (since $T > T_M$). A shear of $\dot{\gamma} = 1.67 \times 10^{-3}$ was applied to the system along the x-direction. We then equilibrate the system for 5×10^3 time steps before we decrease the temperature to $T = 28°\text{C}$.

Figure 10.8b shows a snapshot of the system at $T = 28°\text{C}$ ($t = 1.08 \times 10^6$), where the hairpins are effectively pushed higher into the solution by the swelling of the gel. Since the hairpins are incompatible with the gel, they are effectively excluded from the gel layer. Hence, as the gel expands, the hairpins are brought into greater proximity to the buoyant particle, enhancing the probability of the catch process. As shown Figure 10.8c, the hairpins arrest the particle at $t = 1.3 \times 10^6$. This "catch" is facilitated by the swelling dynamics of the gel, bringing the hairpins closer to the particle.

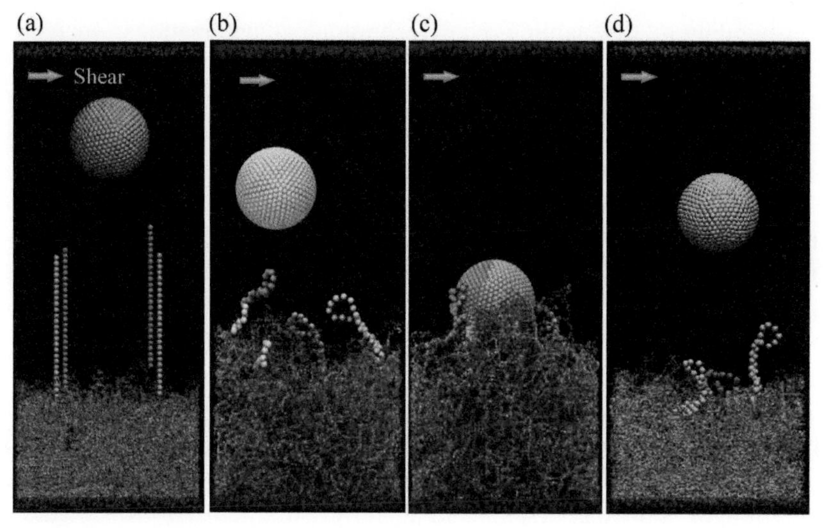

Figure 10.8 (a) Initial morphology of the system with the gel equilibrated at $T = 40°C$ under the shear $\dot{\gamma} = 1.67 \times 10^{-3}$. (b) Snapshot of the system at $t = 1.08 \times 10^6$ after the temperature decreases to $T = 28°C$. (c) Snapshot of the system at $t = 1.3 \times 10^6$. (d) Snapshot of the system encompassing a non-thermo-responsive gel at $t = 5 \times 10^6$ after the system's temperature is decreased to $T = 28°C$, starting with the same initial configuration as in (a).

To elucidate the effect of the thermo-responsive behavior of the gel in our system, we also considered a non-thermo-responsive gel layer by setting the polymer–solvent interaction parameter to a constant value. Consequently, the gel does not response to the temperature change. As shown in Figure 10.8d, when the system is cooled from $T = 40°C$ to $28°C$, the fibers form hairpins, but the gel does not expand. Here, the average height of the gel–fluid interface is 10 units lower than that in the case involving the thermo-responsive gel. Therefore, the hairpins can explore only a limited region and no catch is observed even after 5×10^6 time steps. Hence, our results show that the swelling of the gel facilities the catch process.

10.6 Conclusions

With the aid of computational modeling, our goal was to design artificial cilial layers that could be "programmed" to perform a variety of useful functions within microfluidic devices. The functions ranged from mimicking fundamental biological processes, such as signaling, chemotaxis, and self-regulation, to enabling vital technological applications, such as inhibiting the fouling of the instrument and the selective removal of particles in solution. To controllably and repeatedly actuate these artificial cilia, we devised systems that integrated the properties of stimuli-responsive materials and the appropriate environmental cues. We then used computational approaches to establish means to maximize the synergy between the materials' properties and these cues, and thus enable the artificial system to display the desired dynamics.

In the majority of the systems considered here, we utilized the responsive behavior of polymer gels to drive the dynamic reconfiguration of the system. The chemo-responsive BZ gels both generated and responded to concentration gradients of the activator, u, in the solution. In effect, the BZ cilia propagated signals among themselves through the generation and diffusion of u in the fluid. We showed that the cilia bent toward the highest concentration of u in the system and thereby exhibited a form of auto-chemotaxis, moving in response to chemo-attractants that the system itself produced. This collective behavior can be controlled with light and could be used to shuttle particles from one cilium to another, and thus controllably transport cargo in a microchamber.

We modeled another polymer gel-based system where the synergistic interactions among the different components lead to the oscillations of the embedded cilia, which formed a self-regulating device. With the tips of the cilia extending into the upper fluid, the catalysts on the tips and the reagents in the fluid underwent an exothermic reaction, which caused the underlying LCST gel to collapse. The collapse of the gel shifted the position of the cilial tips so that they no longer were localized in the reagent-laden fluid; in essence, the contraction of the gel served as an "off" switch that terminated the reaction. With the heat dissipated in the system, the gel cooled down and expanded, thereby bringing the cilia into contact with the upper fluid and turning the reaction back "on". This level of self-regulation was accomplished by autonomous interlocking interactions in the multi-component system and thus, resembles the homeostatic activity of biological systems.

In addition to utilizing the thermo-responsive behavior of LCST gels, we functionalized the embedded cilia with spirobenzopyran chromophores, which extended into the bulk of the gel, and allowed us to use light to regulate the dynamic behavior of the system. We expressly concentrated on functionalization with spirobenzopyran chromophores because the temperature of the sample remains constant during the light-induced collapse of the gel [29]. Namely, in this system, illumination does not lead to a heating of the sample [29]; this behavior allowed us to exploit the localized shrinking that occurs only around the functionalized fibers when the sample is illuminated. Hence, by tailoring the arrangement of these fibers within the gel, we can regulate the functionality of the system, designing cilial layers that display the dexterity of hand-like movement and thus, could be used as robotic manipulators.

An imposed flow field also provides an effective stimulus to control the dynamic behavior of the artificial cilia. We considered a scenario where the imposed flow enabled rod-like cilia to selectively trap and bind particles in the upper solution and transport these particles into the lower flowing fluid, from where they could be extracted from the system. Hence, the device could be used for filtration or purification applications. We also introduced a flow field in a system where thermo-responsive artificial cilia were bound to thermo-responsive gels. The concerted response of the thermo-responsive components to variations in temperature allowed the system to trap particles at low temperatures and release them at elevated temperatures. The imposed flow enabled the released particles to be readily removed from the underlying surface and thus, prevents the fouling of the system.

An important goal for future studies is to expand the range of multi-component coatings where cooperative interactions among the components lead to the dynamical properties of biological cilia. Notably, the movement of cilia on marine organisms plays a vital role in inhibiting the biofouling of the organism's surface. Biofouling of submerged man-made surfaces is a critical and costly problem at a range of length scales, from the walls of microfluidic chambers to the hulls of ships. Of particular technological importance is designing artificial cilia that provide purely physical routes to preventing the attachment of fouling agents and enabling their easy removal. Fundamental studies in this area are also needed to elucidate the dependence of biofouling on the architectural, chemical, and dynamic features of artificial cilial coatings. Ultimately, such studies will facilitate the development of novel materials for new, environment-friendly solutions to the general problems of biofouling.

References

[1] Dayal P, Kuksenok O, Bhattacharya A, et al. Chemically-mediated communication in self-oscillating, biomimetic cilia. *J. Mater. Chem.*, 22(1):241–250, 2012.

[2] Yashin VV, and Balazs AC. Theoretical and computational modeling of self-oscillating polymer gels. *J. Chem. Phys.,* 126(12):124707, 2007.

[3] Yashin VV, Kuksenok O, and Balazs AC. Modeling autonomously oscillating chemo-responsive gels. *Prog. Polym. Sci.*, 35(1–2):155–173, 2010.

[4] Kuksenok O, Yashin VV, and Balazs AC. Three-dimensional model for chemoresponsive polymer gels undergoing the Belousov-Zhabotinsky reaction. *Phys. Rev. E*, 78(4):041406, 2008.

[5] He X, Aizenberg M, Kuksenok O, et al. Homeostatic materials with chemo-mechano-chemical self-regulation. *Nature*, 487(7406):214–218, 2012.

[6] Kuksenok O, Balazs, A. C. Stimuli-Responsive Behavior of Composites Integrating Thermo-Responsive Gels with Photo-Responsive Fibers. *Mater. Horiz.*, 3(1):53–62, 2016.

[7] Liu Y, Bhattacharya A, Kuksenok O, et al. Computational modeling of oscillating fins that "catch and release" targeted nanoparticles in Bilayer flows. *Soft Matter*, 12(5):1374–1384, 2016.

[8] Liu Y, Kuksenok O, He X, et al. Harnessing cooperative interactions between thermoresponsive aptamers and gels to trap and release nanoparticles. *ACS Appl. Mater. Interfaces*, 8(44):30475–30483, 2016.

[9] Yoshida R, Takahashi T, Yamaguchi T, et al. Self-oscillating gel. *J. Am. Chem. Soc.,* 118(21):5134–5135, 1996.

[10] Yoshida R, Sakai T, Hara Y, et al. Self-oscillating gel as novel biomimetic materials. *J. Control. Release,* 140(3):186–193, 2009.

[11] Yoshida R, Tanaka M, Onodera S, et al. In-phase synchronization of chemical and mechanical oscillations in self-oscillating gels. *J. Phys. Chem. A,* 104(32):7549–7555, 2000.

[12] Chen IC, Kuksenok O, Yashin VV, et al. Shape- and size-dependent patterns in self-oscillating polymer gels. *Soft Matter,* 7:3141–3146, 2011.

[13] Kuksenok O, Yashin VV, Kinoshita M, et al. Exploiting gradients in cross-link density to control the bending and self-propelled motion of active gels. *J. Mater. Chem.*, 21(23):8360–8371, 2011.

[14] Yuan P, Kuksenok O, Gross DE, et al. UV patternable thin film chemistry for shape and functionally versatile self-oscillating gels. *Soft Matter,* 9:1231–1243, 2013.

[15] Hill TL. *An Introduction to Statistical Thermodynamics.* Addison-Weley: Reading, MA, 1960.

[16] Hirotsu S. Softening of bulk modulus and negative Poisson's Ratio near the volume phase transition of polymer gels. *J. Chem. Phys.*, 94(5):3949, 1991.

[17] Atkin RJ, Fox N. *An Introduction to the Theory of Elasticity*. Longman: New York, NY, 1980.

[18] Onuki A. Theory of phase-transition in polymer gels. *Adv. Polym. Sci.*, 109:63–121, 1993.

[19] Barriere B, Leibler L. Kinetics of solvent absorption and permeation through a highly swellable elastomeric network. *J. Polym. Sci. Part B-Polymer Phys.* 41(2):166–182, 2003.

[20] Doi M. Gel dynamics. *J. Phys. Soc. Japan*, 78(5):052001, 2009.

[21] Yashin, V. V, Balazs, A. C. Pattern formation and shape changes in self-oscillating polymer gels. *Science*, 314(5800):798–801, 2006.

[22] Boissonade J. Self-oscillations in chemoresponsive gels: A theoretical approach. *Chaos*, 15(2):23703, 2005.

[23] Roose T, and Fowler AC. Network development in biological gels: Role in lymphatic vessel development. *Bull. Math. Biol.*, 70(6):1772–1789, 2008.

[24] Marrink SJ, de Vries AH, and Mark AE. Coarse grained model for semiquantitative lipid simulations. *J. Phys. Chem. B*, 108(2):750–760, 2004.

[25] Tabata O, Hirasawa H, Aoki S, et al. Ciliary motion actuator using self-oscillating gel. *Sensors Actuators A - Phys.*, 95(2–3):234–238, 2002.

[26] Szilagyi A, Sumaru K, Sugiura S, et al. Rewritable microrelief formation on photoresponsive hydrogel layers. *Chem. Mater.*, 19(11):2730–2732, 2007.

[27] Satoh T, Sumaru K, Takagi T, et al. Fast-reversible light-driven hydrogels consisting of spirobenzopyran-functionalized poly(N-Isopropylacrylamide). *Soft Matter*, 7(18):8030–8034, 2011.

[28] Satoh T, Sumaru K, Takagi T, et al. Isomerization of spirobenzopyrans bearing electron-donating and electron-withdrawing groups in acidic aqueous solutions. *Phys. Chem. Chem. Phys.*, 13(16):7322–7329, 2011.

[29] Kuksenok O, and Balazs AC. Modeling the photoinduced reconfiguration and directed motion of polymer gels. *Adv. Funct. Mater.*, 23(36):4601–4610, 2013.

[30] Suzuki A, and Tanaka T. Phase-transition in polymer gels induced by visible-light. *Nature*, 346(6282):345–347, 1990.

[31] Suzuki A. Phase-transition in gels of submillimeter size induced by interaction with stimuli. *Adv. Polym. Sci.*, 110:199–240, 1993.

[32] Bray AJ. Theory of phase-ordering kinetics. *Adv. Phys.*, 51(2):481–587, 2002.

[33] Verberg R, Yeomans JM, and Balazs AC. Modeling the flow of fluid/particle mixtures in microchannels: Encapsulating nanoparticles within monodisperse droplets. *J Chem Phys*, 123(22):224706, 2005.

[34] Briant AJ, Papatzacos P, and Yeomans JM. Lattice Boltzmann simulations of contact line motion in a liquid-gas system. *Philos Trans A Math Phys Eng Sci.*, 360(1792):485–495, 2002.

[35] Briant A, and Yeomans J. Lattice Boltzmann simulations of contact line motion. II. Binary fluids. *Phys. Rev. E*, 69(3):031603, 2004.

[36] de Gennes P. Wetting: Statics and dynamics. *Rev. Mod. Phys.*, 57(3):827–863, 1985.

[37] Alexeev A, Verberg R, and Balazs AC. Modeling the motion of microcapsules on compliant polymeric surfaces. *Macromolecules*, 38(24):10244–10260, 2005.

[38] Bhattacharya A, and Balazs AC. Stiffness-modulated motion of soft microscopic particles over active adhesive cilia. *Soft Matter*, 9(15):3945, 2013.

[39] Succi S. *The Lattice Boltzmann Equation: For Fluid Dynamics and beyond*; Oxford University Press, 2001.

[40] Swift M, Orlandini E, Osborn W, et al. Lattice Boltzmann simulations of liquid-gas and binary fluid systems. *Phys. Rev. E*, 54(5):5041–5052, 1996.

[41] Bouzidi M, Firdaouss M, and Lallemand P. Momentum transfer of a Boltzmann-lattice fluid with boundaries. *Phys. Fluids*, 13(11):3452, 2001.

[42] Cates ME, Stratford K, Adhikari R, et al. Simulating colloid hydrodynamics with lattice Boltzmann methods. *J. Phys. Condens. Matter*, 16(38):S3903–S3915, 2004.

[43] Nguyen NQ, and Ladd A. Lubrication corrections for lattice-Boltzmann simulations of particle suspensions. *Phys. Rev. E*, 66(4):046708, 2002.

[44] Shastri A, McGregor LM, Liu Y, et al. An aptamer-functionalized chemomechanically modulated biomolecule catch-and-release system. *Nat. Chem.*, 7(5):447–454, 2015.

[45] Zhu J, Nguyen T, Pei R, et al. Specific capture and temperature-mediated release of cells in an aptamer-based microfluidic device. *Lab Chip*, 12(18):3504–3513, 2012.

[46] Vishnyakov A, Talaga DS, and Neimark AV. DPD simulation of protein conformations: From alpha-helices to beta-structures. *J. Phys. Chem. Lett.*, 3(21):3081–3087, 2012.

[47] Yong X, Kuksenok O, Matyjaszewski K, et al. Harnessing interfacially-active nanorods to regenerate severed polymer gels. *Nano Lett.*, 13(12):6269–6274, 2013.

[48] Groot RD, and Warren PB. Dissipative particle dynamics: Bridging the gap between atomistic and mesoscopic simulation. *J. Chem. Phys.*, 107(11):4423, 1997.

[49] Hoogerbrugge PJ, and Koelman JMVA. Simulating microscopic hydrodynamic phenomena with dissipative particle dynamics. *Europhys. Lett.*, 19(3):155–160, 1992.

[50] Español P, and Warren P. Statistical mechanics of dissipative particle dynamics. *Europhys. Lett.*, 30(4):191–196, 1995.

[51] Pan W, Fedosov D, Karniadakis G, et al. Hydrodynamic interactions for single dissipative-particle-dynamics particles and their clusters and filaments. *Phys. Rev. E*, 78(4):046706, 2008.

[52] Karatrantos A, Clarke N, Composto RJ, et al. Topological entanglement length in polymer melts and nanocomposites by a DPD polymer model. *Soft Matter*, 9(14):3877, 2013.

[53] Liu Y, McFarlin GT, Yong X, et al. Designing composite coatings that provide a dual defense against fouling. *Langmuir*, 31(27):7524–7532, 2015.

[54] Chang H-Y, Lin Y-L, Sheng Y-J, et al. Multilayered polymersome formed by amphiphilic asymmetric macromolecular brushes. *Macromolecules*, 45(11):4778–4789, 2012.

[55] Groot RD, and Madden TJ. Dynamic simulation of diblock copolymer microphase separation. *J. Chem. Phys.*, 108(20):8713, 1998.

[56] Hellweg T, Dewhurst CD, Brckner E, et al. Colloidal crystals made of poly(N-isopropylacrylamide) microgel particles. *Colloid Polym. Sci.*, 278(10):972–978, 2000.

[57] Guo J, Liang H, and Wang ZG. Coil-to-globule transition by dissipative particle dynamics simulation. *J. Chem. Phys.*, 134(24):244904, 2011.

[58] Soto-Figueroa C, Rodriguez-Hidalgo MD, and Vicente L. Dissipative particle dynamics simulation of the micellization-demicellization process and micellar shuttle of a diblock copolymer in a biphasic system (water/ionic-liquid). *Soft Matter*, 8(6):1871–1877, 2012.

[59] Schild HG. Poly(N-isopropylacrylamide): Experiment, theory and application. *Prog. Polym. Sci.*, 17(2):163–249, 1992.

[60] Sega M, Sbragaglia M, Biferale L, et al. Regularization of the slip length divergence in water nanoflows by inhomogeneities at the Angstrom Scale. *Soft Matter*, 9(35):8526, 2013.

[61] Moseler M. Formation, stability, and breakup of nanojets. *Science*, 289(5482):1165–1169, 2000.

Index

About the Editors

Richard Mayne is a resident Research Fellow in Biocomputing at the University of the West of England (UWE), UK. Having originally qualified and practiced as a Biomedical Scientist, he moved to research and received his PhD in Computer Science in 2016. He specialises in the design and fabrication of biocomputing devices which harness emergence, complexity and cognition in live cells, towards finding new materials and applications for computing technologies, in addition to describing poorly understood natural phenomena in algorithmic terms.

Richard has worked at UWE's Unconventional Computing Laboratory since 2013 and has participated in a number of projects, including developing a range of heterotic, intracellular and morphological slime mould (Physarum) computers, doing non-Boolean collision-based computation with liposomes and reprogramming ciliary dynamics in live cells.

Jaap den Toonder is full professor and Chair of the Microsystems group at Eindhoven University of Technology. He received his Master's degree in Applied Mathematics in 1991 (cum laude), and his PhD degree (cum laude) in 1996, both from Delft University of Technology.

In 1995, he joined Philips Research Laboratories in Eindhoven, The Netherlands, where he worked on a wide variety of applications. In 2008, he became Chief Technologist, leading the R&D programs on (micro-)fluidics and materials science and engineering. Next to his main job at Philips, he was a part-time professor of Microfluidics Technology at Eindhoven University of Technology between 2004 and 2013.

Jaap den Toonder's group focuses its research on the investigation and development of novel microsystems design approaches and out-of-cleanroom fabrication technologies. The application focus is on microfluidic chips, biomedical microdevices, organs-on-chips, and soft microrobotics. The group's research approaches are often biologically inspired, translating principles from nature into technological innovations. Jaap den Toonder has (co-)authored over 100 scientific papers, as well as over 40 patents, and he has given over 50 invited lectures at international conferences.